Kinetic Phenomena and Collective Modes in Highly Anisotropic Organic Conductors

V.G. Peschansky, V.A. Sirenko, D.I. Stepanenko

CSP

Cambridge Scientific Publishers

Physics Reviews
Volume 25, Part 3

Printed in UK

ISBN 978-1-908106-61-2 Paperback

Cambridge Scientific Publishers Ltd
45 Margett Street
Cottenham
Cambridge CB24 8QY
UK
www.cambridgescientificpublishers.com
janie.wardle@cambridgescientificpublishers.com

KINETIC PHENOMENA AND COLLECTIVE MODES IN HIGHLY ANISOTROPIC ORGANIC CONDUCTORS

V. G. Peschansky[(1,2)], V.A. Sirenko[(2)], D. I. Stepanenko[(2)]

[1)] *V.N. Karazin Kharkov National University, 4 Svobody Sq., Kharkov 61077, Ukraine*
[2)] *B. Verkin Institute for Low Temperature Physics and Engineering, National Academy of Sciences of Ukraine,47 Nauki Ave., Kharkov 61103, Ukraine*

Abstract

We present a review of experimental and theoretical studies of transport phenomena in strongly anisotropic organic conductors. Considerable attention is paid to the phenomena that are specific to quasi-2D and quasi-1D conductive structures and have no analogues both in ordinary metals and in truly 2D or 1D conducting systems. Angular magneto-resistance oscillations, de Haas–van Alphen and Shubnikov–de Haas phenomena, high-temperature quantum oscillations of the magnetoresistance, and high-frequency resonances, including those arising due to the movement of electrons on open trajectories, are discussed. The resonant angular oscillations of high-frequency conductivity and weakly damped electromagnetic waves in quasi-2D organic conductors under strong spatial dispersion are considered. We review high frequency collective processes in highly anisotropic conducting systems in the presence of an external magnetic field taking into account the Fermi-liquid interaction. The specific properties of the quasi-two-dimensional electron energy spectrum and of the Fermi-liquid interaction in layered conductors leads to occurrence of weakly-damping eigen oscillations of the electron and spin densities which are absent in a quasi-isotropic metal.

PACS: 74.70.Kn Organic superconductors;
75.15.Gd Galvanomagnetic and other magnetotransport effects;
76.40. + b Diamagnetic and cyclotron resonances;

Keywords: Organic metals, angular oscillations, magnetic breakdown, high-frequency resonances

Contents

Introduction

Highly anisotropic organic conductors have attracted considerable interest in recent decades due to a number of specific properties such as low dimensionality, peculiar behavior in high magnetic fields, existence of the superconducting phase, variety of phase states and the possibility of changing the ground state under comparatively weak external actions, for example, pressure or impurity atom doping. The search for new superconducting materials in the 1960s led to the study of low-dimensional structures of organic origin. Particular attention was given to the synthesis of organic complexes based on tetrathiafulvalene (TTF), bis (ethylenedithio) tetrathiafulvalene (BEDT-TTF), tetraselenatetracene (TST), tetramethyltetraselenafulvalene (TMTSF) with different anions X, which would have the filamentary crystallographic structure and relatively high electrical conductivity in a single direction. By 1981, more than 400 organic quasi-one-dimensional (Q1D) conductors had been synthesized, although only a small number exhibited a superconducting transition temperature T_c, which was below or about 1 K [1–3]. More promising in this respect were layered organic charge-transfer complexes with quasi-two-dimensional electron energy spectrum. The first quasi-two-dimensional organic superconductor β-(BEDT-TTF)$_2$I$_3$ was obtained in 1984 in Chernogolovka [4]. At atmospheric pressure, its superconducting transition temperature T_c was 1.5 K, and under an external pressure of about 1 kbar, T_c increased to 8 K. It has been found that many (BEDT-TTF)$_2$X charge-transfer complexes based on tetrathiafulvalene with different anions X exhibit a well pronounced metallic conductivity. Their electrical resistance decreases with decreasing temperature and vanishes at $T = T_c$, which is significantly different for different crystallographic modifications. For example, at atmospheric pressure, a conductor κ-(BEDT-TTF)$_2$I$_3$ exhibits the superconducting transition at $T_c = 3.6$ K, while in a charge transfer complex κ-(BEDT-TTF)$_2$Cu[N(CN)$_2$]Cl at a low pressure of 300 bar, T_c is equal to 12.8 K. Higher superconducting transition temperatures have been found in organic conductors based on fullerene C_{60} (up to 40 K) and in metal oxide compounds.

Improvements in the preparation technology of organic conductors yielded quite perfect single-crystal samples, in which an electron in a magnetic field **H** of 10 T at low temperatures is able to make a few turns on the cyclotron orbit with a frequency ω_H during the mean free time τ. In 1988, in Schegolev's laboratory (Institute of Solid State Physics, Academy of Sciences of the USSR) in a magnetic field up to 14 T, the Shubnikov-de Haas magnetoresistance oscillations were observed in a conductor β-(BEDT-TTF)$_2$IBr$_2$ [5, 6]. Shortly after, the Shubnikov-de Haas effect was also detected in organic charge-transfer complexes (BEDT-TTF)$_2$AuI$_2$ [7], (BEDT-TTF)$_2$Cu(NCS)$_2$, [8, 9] (BEDT-TTF)$_2$I$_3$ [9–11]. Observation of this quantum oscillation effect has indicated that electrical conductivity in organic conductors is due to a group of fermions, which are similar to the conduction electrons in ordinary metals, and the condition $\omega_H \tau \gg 1$, which is necessary to reconstruct their energy spectrum from experimental data on the magnetization and magnetoresistance in a strong magnetic field, can be realized.

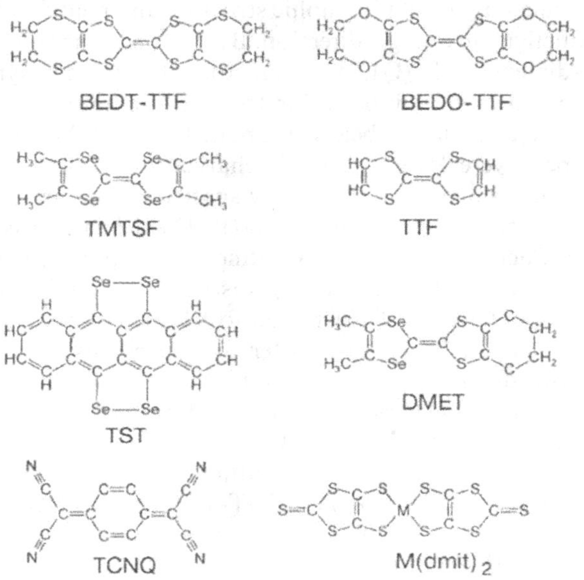

Fig. 1. Examples of organic molecules forming the basis of some organic conductors.

A characteristic feature of the electronic properties of organic metals is a pronounced quasi-1D or quasi-2D anisotropy arising due to their crystal structure. The main structural elements of these compounds are organic molecules or molecular complexes, having donor or acceptor properties – (TTF), (BEDT-TTF), (TMTSF), (TST), and others. Examples of the organic molecules constituting the basis of some organic conductors are shown in Fig. 1.

Ion radicals of organic molecules form regular stacks along the specific direction. Conductivity along the stacks is several orders higher than the electrical conductivity in the transverse direction. As an example, the crystal structure of Q1D conductors (TMTSF)PF_6 at room temperatures and ambient pressure is shown in Fig. 2, [12]. In several conducting complexes (e.g., salts of BEDT-TTF) organic molecules do not form separate stacks but whole conductive layers alternating with layers of anions. The distance between the layers is much greater than the interatomic distances within the layer. The overlap of the electron wave functions belonging to different layers is

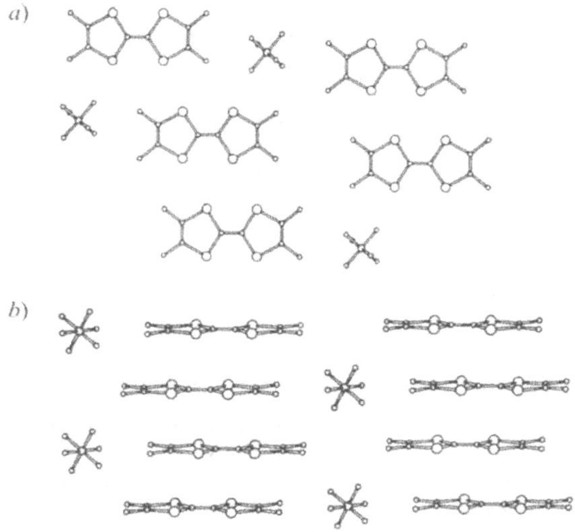

Fig. 2. Structure of (TMTSF)PF_6. The organic molecules (TMTSF) form stacks along the a axis; the anions PF_6^- are located between the stacks, [12]. Crystallographic data at room temperature: $a = 7.297$ Å, $b = 7.711$ Å, $c = 13.522$ Å, $\alpha = 83.39°$, $\beta = 86.27°$, $\gamma = 71.01°$. View along the a direction (a); Side-view of stacks (tilted 10°).

small and as a result, the energy of the conduction electrons depends weakly on the momentum projection on the normal to the layers. Bis (ethylenedithio) tetrathiafulvalene forms the basis of the best-known organic metals. BEDT-TTF cation-radicals are packed in layers separated by the sheets of anion molecules. Within the layers, BEDT-TTF molecules are in close proximity to each other, leading to a significant overlap of the molecular orbitals. As a result, the electrons are free to move from molecule to molecule within the BEDT-TTF plane, while in the direction perpendicular to the BEDT-TTF plane, the organic molecules are separated from each other and the probability of electron transition from one layer to another is small. Thus, a layered structure with a strong anisotropy of the electrical conductivity is formed. Organic conductors are commonly classified according to their crystal structure, and the basic mechanisms of arrangement of the organic molecules are denoted as α, β, β'', δ, χ, κ, λ and θ phases. Each type of packing leads to distinctive features in the Fermi surface topology. Figure 3(a) shows the crystal structure of

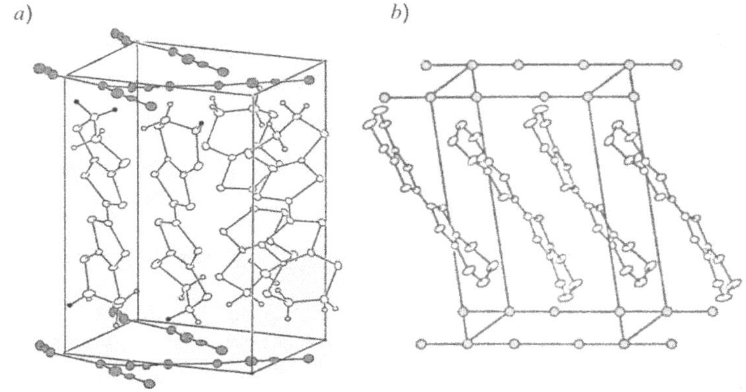

Fig. 3. a) Structure of κ-(BEDT-TTF)$_2$Cu(NCS)$_2$: the (BEDT-TTF)$_2$ molecules are packed in planes, separated by the layers of Cu(NCS)$_2$ anions, [13]. Crystallographic data at room temperature: $a = 16.256$ Å, $b = 8.456$ Å, $c = 13.143$ Å, $\alpha = 90°$, $\beta = 110.28°$, $\gamma = 90°$. b) Structure of a quasi-two-dimensional conductor β-(BEDT-TTF)$_2$I$_3$, [14]. Crystallographic data at room temperature: $a = 6.615$ Å, $b = 9.100$ Å, $c = 15.286$ Å, $\alpha = 94.38°$, $\beta = 95.59°$, $\gamma = 109.78°$.

the κ-phase of a quasi-two-dimensional conductor (BEDT-TTF)$_2$Cu(NCS)$_2$ [13]. The structure of another quasi-two-dimensional conductor β-(BEDT-TTF)$_2$I$_3$ with a different arrangement of organic molecules is shown in Fig. 3(b) [14]. As in the case of κ-(BEDT-TTF)$_2$Cu(NCS)$_2$, BEDT-TTF molecules are packaged in layers separated by the sheets of anion, in this case, I$_3$. A similar structure is exhibited by the β-(BEDT-TTF)$_2$X family with other anions X [15, 16].

Typically, high quality single crystals of organic metals are produced by electrolysis. During this process, the organic molecules are oxidized electrochemically, thereby forming conductive salts. Electrocrystallisation is a slow process, which takes from one week to several months. The minimum current density of about 1 mA/cm^2 required for the crystal growth is maintained in the electrolyte. Typically, Q1D materials have an acicular shape with a length of several millimeters, while Q2D salts from plates with the side lengths of few millimeters and the thickness a few tenths of a millimeter.

The Fermi surface (FS) $\varepsilon(\mathbf{p}) = \varepsilon_F$ of layered conductors is open and slightly corrugated along the momentum projection $p_z = \mathbf{pn}$, where $\mathbf{n}$ is the normal to the layers. It can be constructed using simple topological

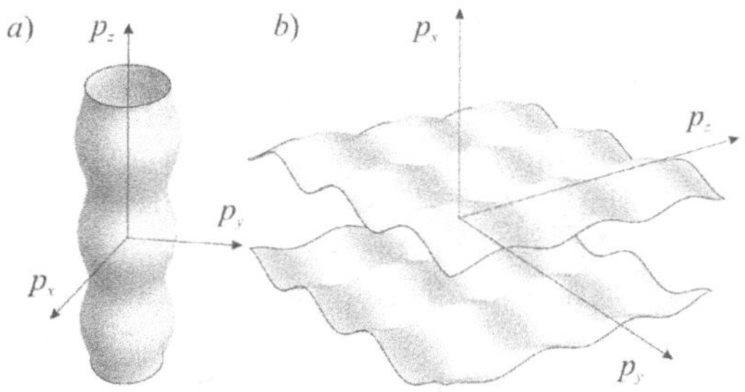

Fig. 4. Elements of the Fermi surface: a) Q2D cylinder, b) Q1D plane.

elements in the shape of slightly corrugated cylinders and slightly corrugated surfaces, see Fig. 4. Detection of the quantum Shubnikov-de Haas oscillations of magnetoresistance in almost all currently investigated layered organic conductors based on BEDT-TTF salts indicates that at least one of the Fermi surface sheets is a slightly corrugated cylinder since the carriers in a corrugated planar Fermi surface sheet do not contribute to the formation of quantum oscillation effects.

In quasi-two-dimensional conductors, plane cross sections of the FS are almost identical at different $p_H = (\mathbf{pH})/H$ to the extent of the smallness of the two-dimensionality parameter η of the energy spectrum of charge carriers, which is the ratio of maximum velocity of electrons across the layers to the characteristic Fermi velocity of the electrons along the layers v_F. Yamaji [17] has shown, using a very simple model of the electron energy spectrum, that for certain orientations of the magnetic field with respect to the layers, the cross-sectional area of electron orbits $S(p_H)$ is independent of p_H in the linear approximation with respect to η. This explains a periodic increase in the amplitude of the Shubnikov-de Haas oscillations when varying $\tan\vartheta$. The theory of the angular magnetoresistance oscillations in layered conductors for an electron energy spectrum of an arbitrary shape has been developed in Ref. 18. From the periods of these oscillations, the diameters of the FS in the form of slightly corrugated cylinders and even the corrugation extent of the planar FS sheets can be determined [19].

Under the action of an external perturbation, such as pressure, the topological structure of the FS can exhibit significant changes, which are accompanied by the anomalous behavior of the thermodynamic and kinetic characteristics of the charge carrier system in degenerate conductors. This electronic topological transition predicted by Lifshitz in 1960, [20] was soon experimentally discovered and investigated in detail in many metals and alloys in the normal and superconducting states, and, during the last 30 years, in metal–dielectric films (MDF), nanostructures, and a variety of low-dimensional conductors. In layered conductors with a multisheet Fermi surface, a certain convergence of the FS sheets under pressure makes it possible for the conduction electrons to move from one sheet to another due to the magnetic breakdown. In this case, the quasi-classical trajectory of the charge carriers becomes complicated and entangled. The anomalous behavior of the electronic characteristics under these conditions is actively studied both theoretically and experimentally.

This paper presents an overview of the transport phenomena in organic conductors in high magnetic fields. Considerable attention is paid to the phenomena characteristic only for quasi-1D and quasi-2D structures, arising due to the peculiarities of the topological structure of the FS.

1. Research techniques of kinetic processes in the conducting systems

The most general description of the system of the conduction electrons with a dispersion relation $\varepsilon(\mathbf{p})$ is the description by means of statistical operator $\hat{\mathsf{R}}$ that satisfies the Liouville-Neumann equation

$$i\hbar\frac{\partial\hat{\mathsf{R}}(t)}{\partial t} = [\hat{\mathsf{H}}, \hat{\mathsf{R}}(t)]. \qquad (1.1)$$

The Hamiltonian for the system $\hat{\mathsf{H}} = \hat{\mathsf{H}}_0 + \hat{V}$ can be conveniently represented as the sum of the unperturbed Hamiltonian $\hat{\mathsf{H}}_0$ and interaction operator $\hat{V}$ of the charged particles with the electromagnetic field. The solution of this equation with initial condition

$$\hat{\mathsf{R}}(-\infty) = \hat{\mathsf{R}}_0 = \exp\left\{\frac{\Omega}{T} - \frac{1}{T}(\hat{\mathsf{H}}_0 - \mu\hat{\mathsf{N}})\right\} \qquad (1.2)$$

allows us to find the average of current density

$$\mathbf{J} = \mathrm{Tr}(\hat{\mathbf{J}}\hat{\mathsf{R}}) \qquad (1.3)$$

and other physical quantities. Here $\hat{\mathsf{R}}_0$ is the Gibbs equilibrium statistical operator, Ω is the thermodynamic potential, $\hat{\mathsf{N}}$ is the operator of particles number, μ is the chemical potential, T is the temperature,

$$\hat{\mathbf{J}}(\mathbf{r}) = \frac{e}{2}\sum_n \{\hat{\mathbf{v}}(\hat{\mathbf{p}}_n)\delta(\mathbf{r}-\mathbf{r}_n) + \delta(\mathbf{r}-\mathbf{r}_n)\hat{\mathbf{v}}(\hat{\mathbf{p}}_n)\} + c\,\mathrm{rot}\sum_n \hat{\boldsymbol{\mu}}_{0n}\delta(\mathbf{r}-\mathbf{r}_n) \tag{1.4}$$

is the operator for current density, $\hat{\mathbf{v}} = \partial\hat{\varepsilon}/\partial\hat{\mathbf{p}}$ is the operator for velocity, $\hat{\mathbf{p}}_n = -i\hbar\partial/\partial\mathbf{r}_n - e\mathbf{A}(\mathbf{r}_n, t)/c$ is the operator for the kinematic momentum, $\mathbf{A}(\mathbf{r}, t) = \mathbf{A}_0(\mathbf{r}) + \mathbf{A}^{\sim}(\mathbf{r}, t)$, $\mathbf{A}_0(\mathbf{r})$ and $\mathbf{A}^{\sim}(\mathbf{r}, t)$ are vector potentials for constant homogeneous and alternating magnetic fields, correspondingly, $\hat{\boldsymbol{\mu}}_0$ is the operator for magnetic moment of the electron, and the summation with respect to n is over all electrons of the system.

In the interaction representation, the equation for the statistical operator $\hat{\tilde{\mathsf{R}}}(t) = \exp(i\hat{\mathsf{H}}_0 t/\hbar)\hat{\mathsf{R}}(t)\exp(-i\hat{\mathsf{H}}_0 t/\hbar)$ can be written as

$$i\hbar\frac{\partial\hat{\tilde{\mathsf{R}}}(t)}{\partial t} = [\hat{\tilde{V}}(t), \hat{\tilde{\mathsf{R}}}(t)], \quad \hat{\tilde{V}}(t) = \exp(i\hat{\mathsf{H}}_0 t/\hbar)\,\hat{V}\exp(-i\hat{\mathsf{H}}_0 t/\hbar). \tag{1.5}$$

Using Eq. (1.5) one can easily find the following integral equation for the operator $\hat{\tilde{\mathsf{R}}}$

$$\hat{\tilde{\mathsf{R}}}(t) = \hat{\tilde{\mathsf{R}}}_0 - \frac{i}{\hbar}\int_{-\infty}^{t} dt'[\hat{\tilde{V}}(t'), \hat{\tilde{\mathsf{R}}}(t')]. \tag{1.6}$$

Considering the interaction of the system with a weak electromagnetic field, we can expand the statistical operator in powers of $\hat{\tilde{V}}(t)$:

$$\hat{\tilde{\mathsf{R}}}(t) = \sum_{n=0}^{\infty}\hat{\tilde{\mathsf{R}}}_n(t). \tag{1.7}$$

In a linear approximation with respect to the weak electromagnetic field, the Hamiltonian of interaction has the form

$$\hat{V}(t) = -\frac{1}{c}\int d^3\mathbf{r}\mathbf{A}^{\sim}(\mathbf{r}, t)\hat{\mathbf{I}}(\mathbf{r}) + \int d^3\mathbf{r}\Phi(\mathbf{r}, t)\hat{\rho}(\mathbf{r}), \tag{1.8}$$

where the operator of current density of unperturbed system $\hat{\mathbf{I}}(\mathbf{r})$ is determined by setting $\mathbf{A}(\mathbf{r}, t) = \mathbf{A}_0(\mathbf{r})$ in Eq. (1.4) and $\Phi(\mathbf{r}, t)$ is the scalar potential of the electromagnetic field, $\hat{\rho}(\mathbf{r})$ is the operator for electric charge density

$$\hat{\rho}(\mathbf{r}) = e\sum_n \delta(\mathbf{r}-\mathbf{r}_n). \tag{1.9}$$

Substituting (1.8) and (1.7) in (1.6), we obtain

$$\hat{\hat{\mathsf{R}}}_1 = \frac{i}{\hbar}\int_{-\infty}^{t} dt' \int d^3\mathbf{r}' \left(-\frac{1}{c}\mathbf{A}^{\sim}(\mathbf{r}',t)[\hat{\hat{\mathbf{I}}}(\mathbf{r}',t),\hat{\mathsf{R}}_0] + \Phi(\mathbf{r},t)t[\hat{\hat{\rho}}(\mathbf{r}',t),\hat{\mathsf{R}}_0\right). \tag{1.10}$$

The current density in the first order in $\mathbf{A}^{\sim}(\mathbf{r},t)$ is given by the following expression [21, 22]

$$J_i(\mathbf{r}) = \mathrm{Tr}(\hat{J}_i\hat{\mathsf{R}}_0) + \frac{i}{\hbar}\int_{-\infty}^{t} dt' \int d^3\mathbf{r}' \left(-\frac{1}{c}A_k^{\sim}(\mathbf{r}',t)\mathrm{Tr}\{\hat{\mathsf{R}}_0[\hat{\hat{I}}_k(\mathbf{r}',t),\hat{\hat{I}}_i(\mathbf{r},t)]\} \right.$$
$$\left. + \Phi(\mathbf{r},t)\mathrm{Tr}\{\hat{\mathsf{R}}_0[\hat{\hat{\rho}}(\mathbf{r}',t),\hat{\hat{I}}_i(\mathbf{r},t)]\}\right), \tag{1.11}$$

here

$$\hat{\hat{\mathbf{I}}}(\mathbf{r},t) = \exp(i\hat{\mathsf{H}}_0 t/\hbar)\hat{\mathbf{I}}\exp(-i\hat{\mathsf{H}}_0 t/\hbar)$$

and

$$\hat{\hat{\rho}}(\mathbf{r},t) = \exp(i\hat{\mathsf{H}}_0 t/\hbar)\hat{\rho}\exp(-i\hat{\mathsf{H}}_0 t/\hbar)$$

are operators $\hat{\mathbf{I}}$ and $\hat{\rho}$ in the interaction representation. Using the operator equality

$$[\hat{I}, e^{-\hat{\mathsf{H}}/T}] = -e^{-\hat{\mathsf{H}}/T}\int_0^{1/T} d\zeta\, e^{\zeta\hat{\mathsf{H}}}[\hat{I},\hat{\mathsf{H}}]e^{-\zeta\hat{\mathsf{H}}}$$

Eq. (1.11) can be transformed to the form [23]

$$\mathbf{J} = \mathbf{j} + \mathbf{j}^{(\mathbf{m})}, \tag{1.12}$$

$$j_i(\mathbf{r},t) = \int_{-\infty}^{t} dt' \int d^3\mathbf{r}' Q_{ik}(\mathbf{r},\mathbf{r}',t-t')E_k(\mathbf{r}',t'), \tag{1.13}$$

$$j_i^{(\mathbf{m})}(\mathbf{r},t) = \frac{1}{c}\int d^3\mathbf{r}' Q_{ik}(\mathbf{r},\mathbf{r}',0)[A_k^{\sim}(\mathbf{r}',t) - A_k^{\sim}(\mathbf{r},t)], \tag{1.14}$$

where

$$Q_{ik}(\mathbf{r},\mathbf{r}',t-t') = \int_0^{1/T} d\zeta \operatorname{Tr}\{\hat{\mathsf{R}}_0 \hat{\hat{I}}_k(\mathbf{r}',t'-i\hbar\zeta)\hat{\hat{I}}_i(\mathbf{r},t)\}. \tag{1.15}$$

When the perturbation of vector potential does not depend on time and in the absence of an electric field, the total current density of electron system is equal to $\mathbf{j}^{(\mathbf{m})}$. In this case formula (1.14) can be interpreted as the magnetization current density (i.e. inducted by magnetic field) and may be used for calculation of a magnetic susceptibility [24]. For a time dependent electromagnetic field, strictly speaking, it is impossible to separate current density into the conduction current and current of magnetization as formula (1.13) includes the terms proportional to the vortical electric field and, therefore, proportional to alternating magnetic field.

In the homogeneous medium the kernel of the integral operator (1.15) depends only on difference $\mathbf{r}-\mathbf{r}'$

$$Q_{ik}(\mathbf{r},\mathbf{r}',t-t') = Q_{ik}(\mathbf{r}-\mathbf{r}',0,t-t'). \tag{1.16}$$

Substituting in the formula (1.15) the Fourier components of the current density and the electric field $\{\mathbf{j}(\mathbf{r},t),\mathbf{E}(\mathbf{r},t)\} = \{\mathbf{j}(\omega,\mathbf{k}),\mathbf{E}(\omega,\mathbf{k})\}\exp(i\mathbf{kr}-i\omega t)$, and passing to the new variables $\mathbf{r}''=\mathbf{r}-\mathbf{r}'$ and $t''=t-t'$, it is easy to find the conductivity tensor

$$\sigma_{ik}(\omega,\mathbf{k}) = \int_0^{\infty} dt'' e^{i\omega t''} \int d^3\mathbf{r}'' e^{-i\mathbf{kr}''} \int_0^{1/T} d\zeta \operatorname{Tr}\{\hat{\mathsf{R}}_0 \hat{\hat{I}}_k(0,-i\hbar\zeta)\hat{\hat{I}}_i(\mathbf{r}'',t'')\}. \tag{1.17}$$

Formulas (1.13)–(1.17) are written in the many-particles formulation, their use for specific calculations is very difficult and possible only at additional simplifying assumptions, for example, in the cases of the small density of particles or weak coupling between particles. Then it is probable to express the kinetic coefficients of a system of interacting particles by means of the wave functions and operators of an ideal gas. Let us consider in more detail the single-particle approach. The operators of physical quantities in this case can be represented as the sum, in which each term contains only variables related to one electron. For free electrons, the constitutive equations can easily be expressed in terms of matrix elements $\langle\alpha|\hat{j}_i(\mathbf{r})|\alpha'\rangle$ of the operator for single-particle current density

$$\hat{\mathbf{j}}(\mathbf{r}) = \frac{e}{2}\left\{\hat{\mathbf{v}}\left(\hat{\mathbf{p}}_1 - \frac{e}{c}\mathbf{A}_0(\mathbf{r}_1)\right)\delta(\mathbf{r}-\mathbf{r}_1) + \delta(\mathbf{r}-\mathbf{r}_1)\hat{\mathbf{v}}\left(\hat{\mathbf{p}}_1 - \frac{e}{c}\mathbf{A}_0(\mathbf{r}_1)\right)\right\} + c\,\mathrm{rot}\boldsymbol{\mu}_0\delta(\mathbf{r}-\mathbf{r}_1) \quad (1.18)$$

in an unperturbed system where α is a set of quantum numbers characterizing the state of the conduction electron.

Let us use representation of secondary quantization for operators of physical quantities. In the absence of interaction, the equilibrium Gibbs statistical operator (1.2) takes the form

$$\hat{\mathsf{R}}_0 = \exp\left\{\frac{\Omega}{T} - \frac{1}{T}\sum_{\alpha}(\varepsilon_\alpha - \mu)\hat{a}^+_\alpha\hat{a}_\alpha\right\}, \quad (1.19)$$

where ε_α is the energy of the conduction electron in the individual state with quantum numbers α, $\hat{a}^+_\alpha$ and $\hat{a}_\alpha$ are ordinary Fermi creation and annihilation operators. Representation α is chosen so that the matrix of the one-particle Hamiltonian $\hat{\varepsilon}(\hat{\mathbf{p}} - (e/c)\mathbf{A}_0(\mathbf{r}))$ will be diagonal.

Substituting the expression for current density in the interaction representation

$$\hat{\hat{\mathbf{I}}}(\mathbf{r}, t) = \sum_{\alpha\alpha'}\exp(i\omega_{\alpha\alpha'}t)\langle\alpha|\hat{j}_i(\mathbf{r})|\alpha'\rangle\hat{a}^+_\alpha\hat{a}_{\alpha'} \quad (1.20)$$

in the formula (1.15) we obtain

$$Q_{ik}(\mathbf{r}, \mathbf{r}', t-t') = \sum_{\alpha_1\alpha_2\alpha_3\alpha_4}\int_0^{1/T} d\zeta\exp\{i\omega_{\alpha_1\alpha_2}(t' - i\hbar\zeta) + i\omega_{\alpha_3\alpha_4}t\} \times\langle\alpha_1|\hat{j}_k(\mathbf{r}')|\alpha_2\rangle\langle\alpha_3|\hat{j}_i(\mathbf{r})|\alpha_4\rangle\mathrm{Tr}\{\hat{\mathsf{R}}_0\hat{a}^+_{\alpha_1}\hat{a}_{\alpha_2}\hat{a}_{\alpha_3}\hat{a}^+_{\alpha_4}\}, \quad (1.21)$$

here $\omega_{\alpha\alpha'} = (\varepsilon_\alpha - \varepsilon_{\alpha'})/\hbar$. Using Wick rules [22], it is simple to find

$$\mathrm{Tr}\{\hat{\mathsf{R}}_0\hat{a}^+_{\alpha_1}\hat{a}_{\alpha_2}\hat{a}_{\alpha_3}\hat{a}^+_{\alpha_4}\} = \delta_{\alpha_2\alpha_3}f^{(0)}_{\alpha_1\alpha_4} + f^{(0)}_{\alpha_1\alpha_2}f^{(0)}_{\alpha_3\alpha_4} - f^{(0)}_{\alpha_2\alpha_3}f^{(0)}_{\alpha_1\alpha_4}, \quad (1.22)$$

where $f^{(0)}_{\alpha_i\alpha_j}$ represent the occupation numbers, i.e. average numbers of particles in the individual quantum state multiplied by the Kronecker delta

$$f^{(0)}_{\alpha_1\alpha_2} = \mathrm{Tr}\{\hat{\mathsf{R}}_0\hat{a}^+_{\alpha_1}\hat{a}_{\alpha_2}\} = \delta_{\alpha_1\alpha_2}f_0(\varepsilon_{\alpha_1}), \quad (1.23)$$

$$f_0(\varepsilon_\alpha) = \left[1 + \exp\frac{(\varepsilon_\alpha - \mu)}{T}\right]^{-1}$$

is the Fermi distribution function. Taking into account Eq. (1.23), formula (1.22) takes the form

$$\begin{aligned}\mathrm{Tr}\{\hat{\mathsf{R}}_0 \hat{a}^+_{\alpha_1}\hat{a}_{\alpha_2}\hat{a}_{\alpha_3}\hat{a}^+_{\alpha_4}\} = {} & \delta_{\alpha_1\alpha_4}\delta_{\alpha_2\alpha_3} f_0(\varepsilon_{\alpha_1})(1 - f_0(\varepsilon_{\alpha_2})) \\ & + \delta_{\alpha_1\alpha_2}\delta_{\alpha_3\alpha_4} f_0(\varepsilon_{\alpha_1}) f_0(\varepsilon_{\alpha_3}). \end{aligned} \tag{1.24}$$

Substituting Eq. (1.24) in the Eq. (1.21) and calculating integral over $d\zeta$, we obtain

$$\begin{aligned}Q_{ik}(\mathbf{r}, \mathbf{r}', t - t') = {} & \sum_{\alpha_1\,\alpha_2} \exp\{i\omega_{\alpha_2\alpha_1}(t - t')\} f_0(\varepsilon_{\alpha_1})(1 - f_0(\varepsilon_{\alpha_2})) \\ & \times \frac{\exp\{T^{-1}(\varepsilon_{\alpha_1} - \varepsilon_{\alpha_2})\} - 1}{\varepsilon_{\alpha_1} - \varepsilon_{\alpha_2}} \langle \alpha_1 | \hat{j}_k(\mathbf{r}') | \alpha_2 \rangle \langle \alpha_2 | \hat{j}_i(\mathbf{r}) | \alpha_1 \rangle. \end{aligned} \tag{1.25}$$

Here we take into account equality to zero the equilibrium current density in unperturbed system

$$\sum_\alpha f_0(\varepsilon_\alpha)\langle \alpha | \hat{j}_i(\mathbf{r}) | \alpha \rangle = 0.$$

By straightforward calculation it is easy to verify the equality

$$f_0(\varepsilon_{\alpha_1})(1 - f_0(\varepsilon_{\alpha_2}))[\exp\{T^{-1}(\varepsilon_{\alpha_1} - \varepsilon_{\alpha_2})\} - 1] = f_0(\varepsilon_{\alpha_2}) - f_0(\varepsilon_{\alpha_1}),$$

whereby, the integral kernel in the Eqs. (1.13), (1.14) in the one-particle approximation can be written as

$$Q_{ik}(\mathbf{r}, \mathbf{r}', t - t') = \sum_{\alpha, \alpha'} e^{i\omega_{\alpha'\alpha}(t - t')} \frac{f_0(\varepsilon_{\alpha'}) - f_0(\varepsilon_\alpha)}{\varepsilon_\alpha - \varepsilon_{\alpha'}} \langle \alpha | \hat{j}_k(\mathbf{r}') | \alpha' \rangle \langle \alpha' | \hat{j}_i(\mathbf{r}) | \alpha \rangle. \tag{1.26}$$

The effects of electron scattering can be taken into account by adding the term $-\tau^{-1}(t - t')$ in the exponent in Eq. (1.26).

Assuming that the space-time dependence of $\mathbf{E}(\mathbf{r}, t)$ and $\mathbf{j}(\mathbf{r}, t)$ has the form $\exp(i\mathbf{kr} - i\omega t)$, we can easily find from Eqs. (1.13), the expression for the conductivity tensor

$$\sigma_{ik}(\omega, \mathbf{k}) = i\sum_{\alpha, \alpha'} \frac{f_0(\varepsilon_{\alpha'}) - f_0(\varepsilon_\alpha)}{\varepsilon_{\alpha'} - \varepsilon_\alpha} \frac{\langle \alpha' | \hat{j}_k(0) | \alpha \rangle \langle \alpha | \hat{\mathrm{J}}_i(\mathbf{k}) | \alpha' \rangle}{\omega_{\alpha'\alpha} - \omega - i\tau^{-1}}, \tag{1.27}$$

where $\langle\alpha|\hat{\mathrm{J}}_i(\mathbf{k})|\alpha'\rangle \equiv \int d^3\mathbf{r}\, e^{-i\mathbf{kr}}\langle\alpha|\hat{j}_i(\mathbf{r})|\alpha'\rangle$. The parameter τ has the significance of a relaxation time in the kinetic equation for the one-particle density matrix with a collision integral in τ – approximation.

To illustrate the application of the formula (1.27), let us consider as a simple example the conductivity of free electron gas in the absence of a magnetic field. In this case the set of quantum numbers characterizing the state of the conduction electron are momentum $\mathbf{p}$ and spin projection σ, $\alpha = \mathbf{p}$, σ. The wave function of an electron represents a plane wave

$$\varphi_\sigma(\mathbf{p}) = \frac{1}{\sqrt{V}}\, e^{i\mathbf{pr}/\hbar}\chi_\sigma,$$

where χ_σ is the spin part of the wave function and V is the volume of the system. Neglecting of the last term in (1.18) proportional to $\hat{\boldsymbol{\mu}}_0$, we have

$$\langle\alpha'|\hat{j}_k(0)|\alpha\rangle\langle\alpha|\hat{\mathrm{J}}_i(\mathbf{k})|\alpha'\rangle = \frac{e^2}{V} v_k(\mathbf{p}) v_i(\mathbf{p}')\delta_{\mathbf{p}',\mathbf{p}+\hbar\mathbf{k}}\delta_{\sigma'\sigma}. \tag{1.28}$$

Substituting Eq. (1.28) in (1.27) and summing the result over $\mathbf{p}'$ and σ, σ' we obtain

$$\sigma_{ik}(\omega, \mathbf{k}) = i\frac{2e^2}{V}\sum_{\mathbf{p}} \frac{f_0(\varepsilon_{\mathbf{p}+\hbar\mathbf{k}}) - f_0(\varepsilon_{\mathbf{p}})}{\varepsilon_{\mathbf{p}+\hbar\mathbf{k}} - \varepsilon_{\mathbf{p}}} \frac{v_i(\mathbf{p}+\hbar\mathbf{k})v_k(\mathbf{p})}{\omega_{\mathbf{p}+\hbar\mathbf{k},\mathbf{p}} - \omega - i\tau^{-1}}. \tag{1.29}$$

Here we denote for brevity $\varepsilon(\mathbf{p}+\hbar\mathbf{k}) = \varepsilon_{\mathbf{p}+\hbar\mathbf{k}}$. In the limiting case $\hbar \to 0$, passing in (1.29) from summation to integration with the help of the relation

$$\frac{1}{V}\sum_{\mathbf{p}} \cdots \to \int \frac{d^3\mathbf{p}}{(2\pi\hbar)^3} \cdots,$$

we obtain well-known quasi-classical formula for the conductivity tensor of a free electron gas

$$\sigma_{ik}(\omega, \mathbf{k}) = -ie^2 \int \frac{2d^3\mathbf{p}}{(2\pi\hbar)^3} \frac{\partial f_0(\varepsilon)}{\partial\varepsilon} \frac{v_i(\mathbf{p})v_k(\mathbf{p})}{\omega - \mathbf{kv} + i\tau^{-1}}. \tag{1.30}$$

In the semi-classical approximation it is more convenient to proceed not from general formulas (1.17), (1.27) but from the quasiclassical Boltzmann equation

$$-i(\omega - \mathbf{kv} + i\tau^{-1})\psi + \frac{e}{c}(\mathbf{v} \times \mathbf{H})\frac{\partial\psi}{\partial\mathbf{p}} = e\mathbf{v}\mathbf{E}(\omega, \mathbf{k}) \tag{1.31}$$

for the nonequilibrium correction to the conduction electron distribution function

$$f(\mathbf{r}, \mathbf{p}, t) = f_0(\varepsilon) - \psi(\omega, \mathbf{k}, \mathbf{p})\frac{\partial f_0(\varepsilon)}{\partial \varepsilon}\exp(-i\omega t + i\mathbf{kr}).$$

The reduction of Eq. (1.1) to the quasiclassical kinetic equation is discussed in more detail in the Section 10.

The electric current density in this case is given by

$$j_i = -e\int \frac{2d^3\mathbf{p}}{(2\pi\hbar)^3}\frac{\partial f_0(\varepsilon)}{\partial \varepsilon}v_i\psi. \tag{1.32}$$

Equation (1.31) can be written as

$$-i(\omega - \mathbf{kv}(t) + i\tau^{-1})\psi(\omega, \mathbf{k}, \mathbf{p}(t)) + \frac{d\psi(\omega, \mathbf{k}, \mathbf{p}(t))}{dt} = e\mathbf{v}(t)\mathbf{E}(\omega, \mathbf{k}), \tag{1.33}$$

where $\mathbf{v}(t)$ and $\mathbf{p}(t)$ are velocity and momentum of an electron, defined from the equations of motion

$$\frac{d\mathbf{p}(t)}{dt} = \frac{e}{c}(\mathbf{v}(t) \times \mathbf{H}), \quad \mathbf{v}(\mathbf{p}) = \frac{\partial \varepsilon(\mathbf{p})}{\partial \mathbf{p}}. \tag{1.34}$$

The solution of the equation (1.33) in an infinite medium has the form

$$\psi = e\int_{-\infty}^{t} dt' v_k(t')E_k \exp\left(i(\omega + i\tau^{-1})(t - t') - \int_{t'}^{t} dt''\mathbf{kv}(t'')\right) \tag{1.35}$$

Substituting (1.35) into (1.32) and passing to new variables

$$d^3\mathbf{p} = \frac{|e|H}{c}d\varepsilon dp_H dt,$$

we obtain a quasi-classical expression for the conductivity tensor in a magnetic field

$$\sigma_{ik}(\omega, \mathbf{k}) = \frac{2|e|^3H}{(2\pi\hbar)^3c}\int d\varepsilon dp_H dt\left(-\frac{\partial f_0(\varepsilon)}{\partial \varepsilon}\right)v_i(t)\int_{-\infty}^{t} dt' v_k(t')$$

$$\times \exp\left(i(\omega + i\tau^{-1})(t - t') - i\int_{t'}^{t} dt''\mathbf{kv}(t'')\right) \tag{1.36}$$

As the variables in the momentum space we have chosen the electron energy ε, the momentum projection $p_H = (\mathbf{pH})/H$ on the magnetic field direction and the time t of motion of an electron in a magnetic field along the trajectory defined by the quasi-classical dynamic equations (1.34). For the phenomena not related to heat transfer, we can replace the derivative of the equilibrium distribution function by the Dirac delta function i.e.

$$\frac{\partial f_0(\varepsilon)}{\partial \varepsilon} = -\delta(\varepsilon - \varepsilon_F).$$

2. Angular resistance oscillations in classically strong magnetic fields

Due to a high anisotropy of the electron energy spectrum, the physical characteristics of organic conductors, in particular, their behavior in a strong magnetic field are substantially different from both the properties of quasi-isotropic metals and those of purely two-dimensional and one-dimensional conducting systems. The low dimensionality of the spectrum of conduction electrons is also manifested in a number of peculiar effects in a strong magnetic field, which are specific only for low-dimensional conductors, namely, the appearance of a series of magnetoresistance peaks upon varying the angle ϑ between the magnetic field vector $\mathbf{H}$ and the normal to the layers $\mathbf{n}$, which are recurring periodically as a function of $\tan\vartheta$. Just as in ordinary metals, magnetotransport phenomena in organic conductors arise when the mean free time of charge carriers s is large enough to exhibit their dynamic properties $\omega_H \tau > 1$.

In the most general form, the energy spectrum of a Q2D conductor can be represented in the form of a rapidly convergent series

$$\varepsilon(\mathbf{p}) = \varepsilon_0(p_x, p_y) + \sum_{n=1}^{\infty} \varepsilon_n(p_x, p_y, \eta) \cos\frac{np_z}{p_0}, \qquad (2.1)$$

where the functions $\varepsilon_n(p_x, p_y, \eta)$ decrease significantly with an increase in their numbers, the greatest of them $\varepsilon_1(p_x, p_y, \eta)$ is of the order of $\eta\varepsilon_F$, and $p_0 = \hbar/a$, a is the distance between elementary crystals cell along p_z-axis.

When a magnetic field $\mathbf{H} = (0, H\sin\vartheta, H\cos\vartheta)$ is significantly tilted from the direction of openness of the FS (p_z-axis) at an angle ϑ, the electron trajectories in momentum space, $\varepsilon(\mathbf{p}) = \varepsilon_F$, $p_H = \mathbf{pH}/H = const$, become strongly elongated along the p_z-axis, and for $\tan\vartheta \gg 1$, it is not possible for an electron to complete a full revolution in its orbit during the mean free time τ. The component velocity v_z

$$v_z = \frac{\partial \varepsilon(\mathbf{p})}{\partial p_z} = -\sum_{n=1}^{\infty} \frac{n}{p_0} \varepsilon_n(p_x, p_y, \eta) \sin \frac{n p_z}{p_0}, \tag{2.2}$$

of charge carriers that move in orbits crossing many unit cells in the momentum space often changes its sign, which leads to a decrease in electrical conductivity in the direction of the opening of the FS. While the period of motion of a conduction electron in a magnetic field $T(\vartheta) = T(0)/\cos\vartheta$ is much smaller than τ, an increase in the resistance with increasing the magnetic field along the z-axis is not yet high. However, for $T(\vartheta) > \tau$, a significant increase with increasing magnetic field begins, and for $\vartheta = \pi/2$, the transverse magnetoresistance increases with the field proportionally to H^2. Averaged over a narrow range of angles near $\vartheta = \pi/2$ (for example, due to the mosaicity of a single-crystal sample), the magnetoresistance linearly increases with increasing a strong magnetic field. Thus, in Reference 25, the possibility of a linear increase in the resistance of many metals with increasing the magnetic field, which was discovered by Kapitsa back in the 1920s [26], has been theoretically confirmed. For $T(\vartheta) \ll \tau$ the electrical conductivity along the z-axis is proportional to the square of the electron drift velocity $\bar{v}_z$. If, at the same time, $\cos\vartheta$ is also negligible, i.e., $T(0)/\tau \ll \cos\vartheta \ll 1$, the main contribution to the average electron velocity $\bar{v}_z$ during the period of its motion in the magnetic field comes from the close vicinities of the turning points of an electron in orbit, where the projection of the momentum p_z takes maximum and minimum values. These contributions can compensate or, on the contrary, strengthen each other if the sign is the same. In metals, the diameter of electron orbits in momentum space, which is equal to $D_z = p_z^{\max} - p_z^{\min}$, essentially depends on the p_H, and the averaging $(\bar{v}_z)^2$ over various cross sections of the FS by the plane $p_H = const$ significantly weakens the non-monotonic dependence of the electrical conductivity on the angle ϑ.

Using the expression for the electron velocity transverse to the layers (2.2) and formula (1.36) at $\omega = 0$, we can write the interlayer conductivity as [18]

$$\sigma_{zz} = \frac{2|e|^3 H}{c(2\pi\hbar)^3 p_0^2} \sum_{n,l=1}^{\infty} nl \int_0^{2\pi p_0 \cos\vartheta} dp_H \int_0^{2\pi/\omega_H} dt \int_{-\infty}^{t} dt' \exp\left(\frac{t'-t}{\tau}\right) \varepsilon_n(p_H, t)$$
$$\times \varepsilon_l(p_H, t') \sin\left(n \frac{p_H - p_y(p_H, t)\sin\vartheta}{p_0 \cos\vartheta}\right)$$
$$\times \sin\left(l \frac{p_H - p_y(p_H, t')\sin\vartheta}{p_0\cos\vartheta}\right). \qquad (2.3)$$

For angles ϑ not too close to $\pi/2$, i.e. for $\eta \tan\vartheta \ll 1$, the functions $p_y(p_H, t)$ and $\varepsilon_n(p_H, t)$ have an asymptotic representation

$$p_y(p_H, t) = p_y(\varphi) + \Delta p_y(p_H, \varphi), \varepsilon_n(p_H, t) = \varepsilon_n(\varphi) + \Delta\varepsilon_n(p_H, \varphi),$$
$$\varphi = \omega_H(p_H)t$$

where $\Delta p_y(p_H, \varphi) \simeq \eta p_y(\varphi)$ and $\Delta\varepsilon_n(p_H, \varphi) \simeq \eta\varepsilon_n(\varphi)$. Substituting these expressions into Eq. (2.3) and integrating with respect to p_H, we obtain

$$\sigma_{zz} = \frac{e^2 m^* \cos\vartheta}{(2\pi)^2 \hbar^3 \omega_H p_0} \sum_{n=1}^{\infty} n^2 \int_0^{2\pi} d\varphi \int_{-\infty}^{\varphi} d\varphi' \exp\left(\frac{\varphi'-\varphi}{\omega_H \tau}\right) \varepsilon_n(\varphi)$$
$$\times \varepsilon_n(\varphi') \cos\left(n \frac{p_y(\varphi) - p_y(\varphi')}{p_0} \tan\vartheta\right), \qquad (2.4)$$

Here m^* is the cyclotron effective mass of electron, $\omega_H = |e|H/(m^* c)$.

Provided that $(\omega_H \tau)^{-1} \gg 1$, the asymptote of Eq. (2.4) has the form

$$\sigma_{zz} = \frac{e^2 \tau m^* \cos\vartheta}{(2\pi)^2 \hbar^3 p_0} \sum_{n=1}^{\infty} n^2 \left(\int_0^{2\pi} d\varphi \varepsilon_n(\varphi) \cos\left(n \frac{p_y(\varphi)}{p_0} \tan\vartheta\right)\right)^2. \qquad (2.5)$$

Equation (2.5) describes the oscillatory dependence of the interlayer conductivity on the tilt angle of the magnetic field. To see this, it suffices to calculate its asymptotic behavior for large values of $\tan\vartheta$. For $\tan\vartheta \gg 1$, the main contribution to the integral with respect to $d\varphi$ comes from the stationary phase points φ_i, which are determined from the equation

$$\frac{dp_y(\varphi_i)}{d\varphi} = -m^* \cos\vartheta \, v_x(\varphi_i) = 0.$$

Assuming that there are two such points and $\varepsilon_n(\varphi_1) = \varepsilon_n(\varphi_2)$, we obtain for the asymptotic expression of the conductivity the following expression

$$\sigma_{zz} = \frac{e^2 \tau \cos^2\vartheta}{2\pi^2\hbar^3 |v_x'(\varphi_1)| \sin\vartheta} \times \sum_{n=1}^{\infty} n^2 \varepsilon_n^2(\varphi_1) \left\{ 1 + \sin\left(\frac{nD_p}{p_0} \tan\vartheta \right) \right\}, \qquad (2.6)$$

where $v_x'(\varphi_1) = \partial v_x(\varphi_1)/\partial\varphi$ and $D_p = p_y(\varphi_2) - p_y(\varphi_1)$ is the cross-section size of the Fermi surface along the y-axis. The terms in the summation in Eq. (2.6) as $\tan\vartheta$ functions, oscillate with a period

$$\mathrm{T}_{\tan\vartheta} = \frac{2\pi p_0}{D_p},$$

taking minimum values equal to zero at the points $\tan\vartheta = 2\pi(l - 1/4)p_0/D_p$, $l = 0, 1, 2, \ldots$.

Equation (2.6) is obtained from Eq. (2.5) assuming that $\tan\vartheta \gg 1$. Evidently, for not too large values of $\tan\vartheta$, all the terms in Eq. (2.5) do not vanish at the same values of ϑ. In this case, the angular dependence of the magnetoresistance is determined by the rate of decrease of the functions $\varepsilon_n(p_x, p_y, \eta)$ with increasing n. Typically, it is sufficient to limit ourselves to the first few terms in the summation (2.5). However, comparing the experimentally observed angular oscillations with the theoretical calculation [27] retaining the four terms in Eq. (2.6) showed that the coefficients decrease more slowly than in the tight-binding approximation, and the genesis of the energy spectrum of charge carriers in layered conductors is more complicated.

The angular magnetoresistance oscillations were first observed experimentally in Q2D metal β-(BEDT-TTF)$_2$IBr$_2$ [5, 6], simultaneously with the Shubnikov-de Haas phenomenon. Later, the phenomenon has been observed in other organic conductors (for example, see the reviews of References [4, 28–36]). The oscillatory dependence of the magnetoresistance on the orientation of the magnetic field is characteristic for almost all organic Q2D conductors. Figure 5 [37] shows an

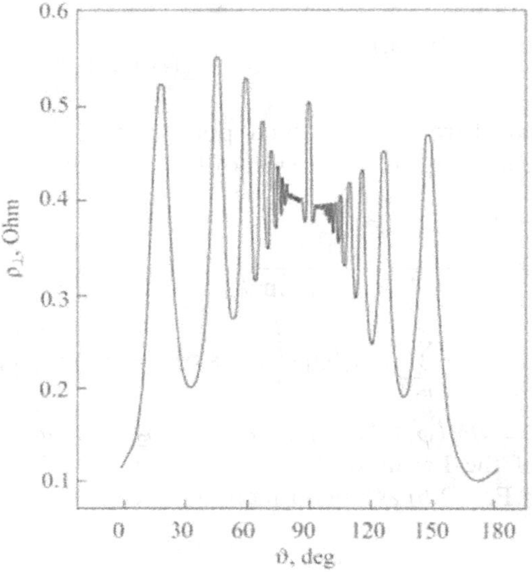

Fig. 5. Dependence of the resistance of a β-(BEDT-TTF)$_2$IBr$_2$ single crystal measured in the direction perpendicular to the highly conductive ab plane in a magnetic field of 15 T at $T = 1.4$ K, on the angle θ between the field direction and the normal to the ab plane [37].

example of such angular dependence of the resistance of a single crystal of a layered conductor β-(BEDT-TTF)$_2$IBr$_2$, measured in the direction perpendicular to the conductive ab-plane upon the rotation of a 15 T magnetic field with respect to the normal to the layers.

The practical significance of this phenomenon lies in the fact that from the periods of angular magnetoresistance oscillations for different planes of rotation of the magnetic field, it is possible to determine the shape of a cross section of the Fermi surface of a quasi-2D conductor. Analysis of experimental data allows us not only to obtain qualitative information on the Fermi surface topology but also the quantitative estimates of its shape and size. Nowadays angular magnetoresistance oscillations are employed in studies of the FS of not only the organic metals, but also other low-dimensional inorganic layered conductors [38–42].

The oscillatory dependence of the resistance on the angle between the magnetic field and the direction of the lowest conductivity also occurs in Q1D conductors with the FS consisting of two slightly

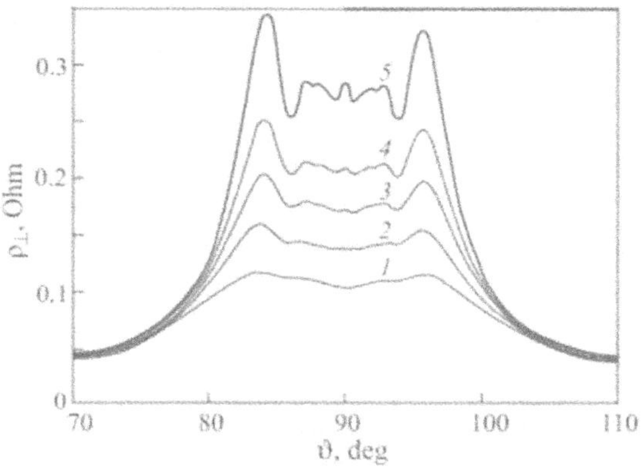

Fig. 6. Angular dependences of the resistance in the c direction of $(TMTSF)_2ClO_4$ measured at $T = 0.5$ K for a magnetic field within the ac-plane[43]. Curves 1–5 correspond to a magnetic field of 3, 4, 5, 6, and 8 T, respectively.

corrugated planes. Figure 6 shows the angular dependence of the resistance of an organic Q1D conductor $(TMTSF)_2ClO_4$ in a magnetic field $\mathbf{H} = (H \sin \vartheta, 0, H \cos \vartheta)$ rotating in the *ac*-plane in the case of current flowing in the direction of the lowest conductivity [43]. Along certain directions of the magnetic field, magnetoresistance peaks were observed.

This effect can be explained by using the equation

$$\varepsilon(\mathbf{p}) = v_F(|p_x| - p_F) + B\cos\frac{p_y}{p_2} + C\cos\frac{p_z}{p_3}, \tag{2.7}$$

for the Q1D energy spectrum of charge carriers, which corresponds to the FS consisting of two slightly corrugated planes. Here $v_F = (A/p_1)\sin(p_F/p_1)$ and p_F are the speed and momentum on the Fermi surface along the direction of maximum conductivity, $A \gg B \gg C$. The characteristic values of the overlap integrals in Q1D organic conductors are of the order of magnitude: 0.5 eV, $B \sim 0.05$ eV, $C \sim 2$ meV. The p_2 and p_3 parameters are determined by the lattice constants: $p_1 = \hbar/a_1$, $p_2 = \hbar/a_2$, $p_3 = \hbar/a_3$, and a_1, a_2, a_3 are the basic periods of a lattice, $\hbar$ is the Planck constant.The dispersion relation (2.7) corresponds to the energy spectrum in the strong coupling approximation linearized near the Fermi level in the direction of the highest conductivity.

In the zero-order approximation in a small parameter $(C/A)\tan\vartheta \ll 1$ the dynamics of electrons in momentum space is determined by the equations

$$p_x = \mathrm{sign}(\bar{p}_x)p_2\frac{v_2}{v_F}\cos\Omega t + \bar{p}_x, \quad p_y = \mathrm{sign}(\bar{p}_x)p_2\Omega t,$$

$$p_z = \frac{p_H}{\cos\vartheta} - p_x\tan\vartheta, \quad v_x = \mathrm{sign}(\bar{p}_x)v_F, \quad v_y = -\mathrm{sign}(\bar{p}_x)v_2\sin\Omega t,$$

$$v_z = v_3\sin\left(\frac{p_H}{p_3\cos\vartheta} - \frac{p_x(t)}{p_3}\tan\vartheta\right). \tag{2.8}$$

Here $\Omega = |e|v_F H\cos\vartheta/cp_2$ – is an analogue of the cyclotron frequency of electrons with an energy spectrum (2.7), $v_2 = B/p_2$ and $v_3 = C/p_3$ are the characteristic charge carrier velocities in the bc-plane, where the values of $\mathrm{sign}(p_x) = \pm 1$ correspond to the different sheets of the Fermi surface, $\bar{p}_x$ is the mean value of the momentum component p_x. For angles ϑ sufficiently close to $\pi/2$, specifically $(C/A)\tan\vartheta \simeq 1$, the motion of electrons becomes significantly more complex, closed trajectories appear due to the weak corrugation of the Fermi surface along the z-axis, and the velocity and momentum components cannot be expressed in terms of elementary functions.

According to Eqs. (2.8) the electron drift velocity in the direction of the lowest conductivity

$$\langle v_z\rangle = \frac{\Omega}{2\pi}\int_0^{2\pi/\Omega} dt\, v_z(t) = v_3 \mathrm{J}_0\left(\frac{v_2}{v_F}\frac{p_2}{p_3}\tan\vartheta\right)\sin\left(\frac{p_H}{p_3\cos\vartheta} - \frac{\bar{p}_x}{p_3}\tan\vartheta\right) \tag{2.9}$$

is an oscillating function of the angle ϑ. Repeating the above arguments set out in relation to Q2D conductors, we arrive at the conclusion that for the directions of a magnetic field for which $\langle v_z\rangle$ is of second order of smallness in $(C/A)\tan\vartheta \simeq 1$, there should be local maxima of the resistance ρ_{zz}. The argument of the Bessel function in Eq. (2.9) includes a small parameter v_2/v_F, therefore the roots of the equation $\langle v_z\rangle = 0$ appear in the range of the tilt angles of a magnetic field, which satisfies the inequality $\tan\vartheta \gg 1$, for which the condition of well-manifested magnetotransport effects $\Omega\tau \gg 1$ is difficult to realize in practice. This explains the fact that the magnetoresistance peaks in Fig. 6 are blurred, and their amplitude is small compared with the amplitude of the resistance oscillations in Q2D conductors (see Fig. 5).

In a magnetic field $\mathbf{H} = (0, H\sin\vartheta, H\cos\vartheta)$ aligned in a plane perpendicular to the conducting chain, an oscillation effect due to the topology of the FS also occurs [44–46]. We write the energy spectrum of the charge carriers using the relation

$$\varepsilon(\mathbf{p}) = v_F(|p_x| - p_F) + \sum_{n,l} A_{nm}\cos\left(n\frac{p_y}{p_2} + l\frac{p_z}{p_3}\right), \tag{2.10}$$

which is a direct generalization of Eq. (2.7). The overlap integrals $A_{n,m}$ decrease rapidly with an increase in the absolute value of the indices n and m and also satisfy the condition of $A_{n,m} = A_{-n,-m}$. The components of the electron velocity can be expressed in terms of elementary functions

$$v_x = \mathrm{sign}(\bar{p}_x)v_F,$$

$$v_y = -\sum_{n,l} n\frac{A_{nl}}{p_2}\sin\left(\mathrm{sign}(\bar{p}_x)\left(n - \frac{p_2}{p_3}l\tan\vartheta\right)\Omega t + \frac{p_H}{p_3\cos\vartheta}\right),$$

$$v_z = -\sum_{n,l} l\frac{A_{nl}}{p_3}\sin\left(\mathrm{sign}(\bar{p}_x)\left(n - \frac{p_2}{p_3}l\tan\vartheta\right)\Omega t + \frac{p_H}{p_3\cos\vartheta}\right). \tag{2.11}$$

Using Eq. (2.11) and formula (1.36) at $\omega = 0$, it is straightforward to find the conductivity components [46]

$$\sigma_{xx} = e^2\nu(\varepsilon_F)v_F^2\tau, \quad \sigma_{xy} = \sigma_{yx} = \sigma_{xz} = \sigma_{zx} = 0,$$

$$\begin{pmatrix} \sigma_{yy} & \sigma_{yz} \\ \sigma_{zy} & \sigma_{zz} \end{pmatrix} = e^2\nu(\varepsilon_F)\tau\sum_{n,l}\begin{pmatrix} n^2\dfrac{A_{nl}^2}{p_2^2} & nl\dfrac{A_{nl}^2}{p_2p_3} \\ nl\dfrac{A_{nl}^2}{p_2p_3} & l^2\dfrac{A_{nl}^2}{p_3^2} \end{pmatrix}$$

$$\times\frac{1}{1 + \left(n - \dfrac{p_2}{p_3}l\tan\vartheta\right)^2\Omega^2\tau^2}, \tag{2.12}$$

where $\nu(\varepsilon_F) = 2p_2p_3/(\pi\hbar^3v_F)$ is the density of electron states at the Fermi level. Local maxima of the conductivity components σ_{ij}, $\{i,j\} = y, z$ may occur for the angles ϑ satisfying the Lebed' resonance condition [47]

$$\tan\vartheta = \frac{n}{l}\frac{p_3}{p_2}. \tag{2.13}$$

This effect is explained as follows. The components of the electron velocity and v_z (Eq. (2.11)) consist of a sum of harmonics $v_{nl}^{(i)}$, $i = y$, z, oscillating with the frequencies

$$\Omega_{nl} = \frac{|e|v_F H}{cp_2}\left(n\cos\vartheta - \frac{p_2}{p_3}\, l\sin\vartheta\right),$$

which are linear combinations of the frequencies Ω and $\Omega_1 = (p_2/p_3)\Omega\tan\vartheta$ with integer coefficients. For the tilt angles of the magnetic field, defined by Eq. (2.13) the ratio Ω/Ω_1 is a rational fraction, and some of the frequencies Ω_{nl} may vanish. The corresponding harmonics $v_{nl}^{(i)}$ are constants, and their contribution to the drift velocity is the highest possible. In other words, the magnetic field ceases to affect them. This direction of the magnetic field corresponds to the maxima of the conductivity components, σ_{ij}, $\{i, j\} = y$, z and the resistance minima in the bc-plane. Since $v_{nl}^{(i)}$ are proportional to the coefficients $A_{n,l}$, rapidly decreasing with increasing n, and l, the resistance minima can only occur at not so high values of n and l. However the minima of the resistance derivative with respect to the angle ϑ are clearly detectable.

The angular resistance oscillations considered in this section are the property of Q1D and Q2D conducting systems. They arise due to the peculiarities of the energy spectrum of the charge carriers, specifically, due to a weak corrugation of the cylinders and planes constituting the topological elements of the FS. Although these phenomena manifest the three-dimensional nature of organic conductors, they are absent in ordinary metals.

3. De Haas-van Alphen effect

Landau's work devoted to the diamagnetism of free electrons [48] has shown for the first time that in a uniform magnetic field **H**, the thermodynamic potential and the magnetization acquire additional terms, rapidly oscillating with changing H^{-1}. Shortly afterwards, the oscillations of conductivity [49] and magnetization [50] have been observed experimentally in bismuth and dubbed the Shubnikov-de Haas (SdH) and de Haas-van Alphen (dHvA) effects, respectively. In subsequent years, these effects have been also observed in other metals. The periods of oscillations have been found to be different in different metals with their value dependent on the orientation of **H** relative to the single-crystal sample. Such results could not be explained in terms of the free electron model. An important step in understanding the nature of the oscillatory phenomena was the work by Onsager, [51] in which he derived the quasi-classical quantization rule for the area $S(\varepsilon, p_H)$ bound by the trajectory of an electron with an arbitrary energy spectrum $\varepsilon(\mathbf{p})$ in momentum space

$$S(\varepsilon, p_H) = \frac{2\pi|e|H}{\hbar c}\left(n + \frac{1}{2}\right), \tag{3.1}$$

and showed that the period of the magnetization oscillations is determined by the extreme value $S_{ext} = S_{ext}(\varepsilon_F, p_H^m)$ of the area $S(\varepsilon_F, p_H)$ (here, p_H^m is projection of the quasi momentum p_H on the direction of an external magnetic field at which $S(\varepsilon, p_H)$ reaches the extremum). Equation defines the allowable electron energy levels in a quantizing magnetic field. A complete theory of quantum oscillations of the magnetization of metals with an arbitrary dispersion law of charge carriers has been developed by Kosevich and Lifshitz. [52] According to Ref. [52], the thermodynamic potential of a system of N conduction electrons in a uniform magnetic field can be written, using the quantization rules (3.1) and the Poisson summation equation, as a sum of two terms: one, which is smoothly dependent on the magnetic field and an oscillating one. Although the oscillating term of the thermodynamic potential $\tilde{\Omega}$ is small compared to the monotonic term $\bar{\Omega}$, at sufficiently low temperatures, it produces the main contribution to the magnetization.

If the temperature is not too low $\lambda = 2\pi^2 T/(\hbar\omega_H) > 1$, the oscillatory part of the magnetization is, in accordance with Ref. [52]

$$\tilde{M} \approx \frac{2M_0}{\pi^{3/2}} \sum_{S_{ext}} \left(\frac{\varepsilon_F}{\hbar\omega_H}\right)^{1/2} \left(\frac{S_{ext}}{m^*\varepsilon_F}\right) \frac{\lambda e^{-\lambda}}{\sqrt{|S''_{ext}|}} \cos\left(\pi\frac{m^*}{m}\right) \times \sin\left(\frac{cS_{ext}}{|e|\hbar H} + \frac{\pi}{4}\operatorname{sign} S''_{ext}\right), \tag{3.2}$$

where the constant $M_0 = \chi_0 H$ and

$$\chi_0 = \frac{1}{(2\pi)^2} \frac{e^2}{m^* c^2} \frac{\sqrt{2m^* \varepsilon_F}}{\hbar}$$

are of the same order of magnitude as the monotonic magnetization term and the absolute value of the Landau diamagnetic susceptibility, respectively;

$$\omega_H = |e|H/(m^* c), \quad m^* = \frac{1}{2\pi} \frac{\partial S_{ext}}{\partial \varepsilon}\bigg|_{\varepsilon=\varepsilon_F}$$

is the cyclotron mass,

$$S''_{ext} = \frac{\partial^2 S_{ext}}{\partial p_H^2}\bigg|_{\varepsilon=\varepsilon_F},$$

and the summation is carried out over all the extreme cross sections of the FS by the plane $p_H = const$. The ratio of the oscillating $\tilde{M}$ and monotonic $\bar{M} \simeq M_0$ magnetization terms are of the order of $(\varepsilon_F/\hbar\omega_H)^{1/2}$. Even greater is the corresponding ratio for the differential magnetic susceptibilities: $\tilde{\chi}/\bar{\chi} \simeq (\varepsilon_F/\hbar\omega_H)^{3/2}$.

Collisions of the charge carriers can be accounted for by introducing a factor $R_D = \exp(-2\pi/(\omega_H \tau)) \equiv \exp(-2\pi^2 T_D/(\hbar\omega_H))$, where τ is the effective relaxation time, into Eq. (3.2) [53]. Due to the scattering of electrons, the Landau levels are broadened, resulting in reduction of the oscillation amplitude, equivalent to what would be observed by increasing the temperature by T_D. The parameter T_D is called the Dingle temperature.

The quantum oscillations of resistance and magnetization – SdH and dHvA phenomena – are most widely used for experimental study of the FS topology in conductive systems. Organic Q2D conductors are a convenient object for studying these phenomena. In 3D metals, the magnetic quantum oscillations occur due to electrons in the extreme sections of the FS by the plane $p_H = const$, while in layered conductors a much larger number of charge carriers are involved in their formation. This is explained by the fact that the cross-sectional area $S(\varepsilon_F, p_H)$ is weakly dependent on p_H, i.e., the dependence $S(\varepsilon_F, p_H)$ on p_H appears only in the first approximation in the two-dimensionality parameters of the FS [17].

If the overlap integral of the wave functions of electrons belonging to different layers is substantially greater than the distance between adjacent Landau levels

$$\eta\varepsilon_F \gg \hbar\omega_H \tag{3.3}$$

(i.e., a significant number of the Landau levels fit into the conduction band in the direction perpendicular to the layers), the dHvA effect is well described by the Lifshitz-Kosevich equation obtained for quasi-isotropic metals [52]. In the case of a slightly corrugated cylindrical Fermi surface, the magnetization oscillations arise due to two extreme sections with similar area – the maximum $S_{\max}$ and minimum $S_{\min}$. Considering only the first harmonic (3.2) in the Lifshitz-Kosevich equation, the magnetization can be written as

$$\begin{aligned} M &\approx M_m\left\{\sin\left(\frac{cS_{\max}}{|e|\hbar H}-\frac{\pi}{4}\right)+\sin\left(\frac{cS_{\min}}{|e|\hbar H}+\frac{\pi}{4}\right)\right\} \\ &= 2M_m\sin\left(\frac{2\pi F}{H}\right)\cos\left(\frac{\pi\Delta F}{H}-\frac{\pi}{4}\right). \end{aligned} \tag{3.4}$$

The amplitude of the magnetization oscillations M_m which is proportional to $(|S''_{ext}|)^{-1/2} \simeq (\eta m^* \varepsilon_F / p_H^{m2})^{-1/2}$, increases with decreasing the two-dimensionality parameter as $1/\sqrt{\eta}$ and significantly exceeds the amplitude of the oscillations in ordinary metals under similar conditions. The fundamental frequency and the beat frequency are

$$F = c(S_{\max}+S_{\min})/(4\pi|e|\hbar), \quad \Delta F = c(S_{\max}-S_{\min})/(2\pi|e|\hbar)$$

and the difference $\Delta S = S_{\max} - S_{\min}$ is proportional to the small parameter η. A typical magnetic field dependence of the magnetization is shown in Fig. 7.

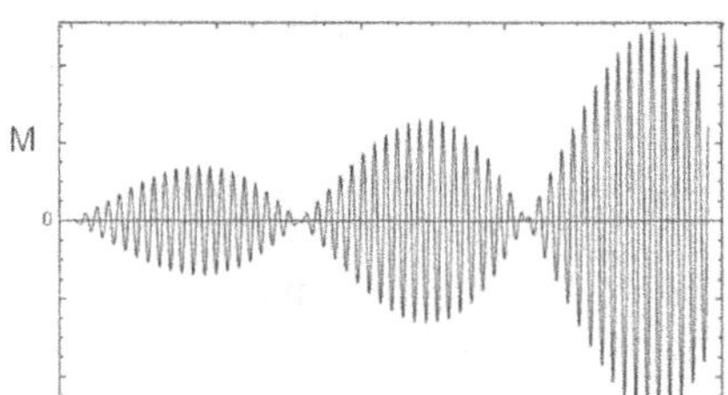

Fig. 7. Characteristic dependence of the magnetization on the magnetic field in a layered Q2D conductor.

From the ratio of the beat and fundamental frequencies, the FS anisotropy parameter can be estimated

$$\eta \simeq \Delta F/F. \tag{3.5}$$

Equation (3.4) is in good agreement with the data obtained from the study of dHvA oscillations in a Q2D organic metal β-(BEDT-TTF)$_2$IBr$_2$, [54]. The amplitude of the fundamental harmonic is modulated with a low frequency determined by the corrugation of the cylindrical FS. Two nodes are clearly observed at 12.5 and 1.7 T. Equation (3.4), can be used to describe the magnetic properties of Q2D organic conductors with the Fermi energy of the order of tenths of eV and the two-dimensionality parameter $\eta \geqslant 10^{-2}$ in magnetic fields $H \sim 10$ T. In the case where the inequality opposite to the condition (3.3) holds, the chemical potential exhibits strong oscillations vs. the reciprocal magnetic field and the quasi-classical approximation based on the Lifshitz-Kosevich theory is not applicable.

Topological peculiarities of the strongly anisotropic FS are manifested not only in the magnetization beats but also in the dependence of the oscillation amplitude of the magnetization and magnetic susceptibility on the magnetic field orientation. The cross sectional area of the FS and the frequency F have a minimum when **H** is perpendicular to the $\vartheta = 0$. When the magnetic field is tilted away from the normal, the electron orbit area increases proportionally to $1/\cos\vartheta$, while the difference ΔS, and, accordingly, ΔF, exhibit an oscillatory dependence of ϑ [55], vanishing (up to the terms of second order in η) for some values of ϑ_i. For these angles, in the linear approximation in η, all the sections of the FS by the plane $p_H = const$ are identical and the amplitude of the oscillations is maximal; the angles ϑ_i correspond to the maxima of the quasiclassical magnetoresistance.

4. Shubnikov-de Haas effect and high-temperature quantum oscillations

An interesting property of layered metals is the existence of low-frequency quantum oscillations (in $1/H$) of conductivity, which are observed at higher temperatures as compared to the oscillations at the fundamental frequency. In 3D conductors, high-temperature quantum oscillations (HTO) have been observed in bismuth and antimony alloys [56–59]. These experiments have been explained in

Ref. [60]. The cause of the HTO in semimetals is interband transitions of the charge carriers belonging to the electron and hole sheets of constant-energy surfaces within the overlap of the valence and conduction bands due to scattering. As a result, harmonics at combined frequencies, which are weakly dependent on the temperature, appear in the oscillating part of the conductivity. The theory of the SdH oscillations requires rigorous accounting of the charge carrier scattering by point defects and other quasi-particles in a quantizing magnetic field. Calculations of the transverse conductivity in a magnetic field $\mathbf{H} = (0, 0, H)$ for a simple parabolic dispersion law of electrons and holes by in a semi-metal $\varepsilon^e(\mathbf{p}) = p^2/2m_e$, $\varepsilon^h(\mathbf{p}) = \varepsilon_{ov} - (\mathbf{p} - \mathbf{p}_0)^2/2m_h$ (where $m_{e,h}$ are the corresponding effective masses and ε_{ov} is the amount of the overlap between the valence and conduction bands), using the Kubo method [21] assuming that elastic scattering by impurities, lead to the following result for the HTO:

$$\sigma_{xx}^{HTO} = \frac{3}{8}\sigma_{mon}\frac{\hbar(\omega_e + \omega_h)}{\varepsilon_{ov}}\sum_{l,l'=1}^{\infty}\frac{(-1)^{l+l'}}{\sqrt{ll'}}R_T\left(\frac{2\pi^2 T}{\hbar\omega_{\overline{ll'}}}\right)\sin\left(\frac{cS_{ll'}^{+}}{\hbar|e|H}\right). \quad (4.1)$$

Here σ_{mon} is the monotonic part of the conductivity, $\omega_{e,h}$ are the cyclotron frequencies of the electrons and holes, $R_T(x) = x/\sinh x$ is the function defining the thermal attenuation,

$$\frac{1}{\omega_{\overline{ll'}}} = \left|\frac{l}{\omega_e} - \frac{l'}{\omega_h}\right|,$$

$S_{ll'}^{+} = lS_e + l'S_h$, $S_{e,h}$ are the areas of the extreme sections of the electron and hole of the FS sheets by the plane $p_H = const$, $\omega_{e,h}$, $T \ll \varsigma_{e,h}$, $\varsigma_e = \varsigma$, $\varsigma_h = \varepsilon_{ov} - \varsigma$ are the electron and hole chemical potentials determined from the electrical neutrality condition.

The main contribution to the HTO is due to the harmonics for which $\omega_{\overline{ll'}}$ is maximal, i.e., $lm_e \approx l'm_h$. The oscillation period of these harmonics is of the order of magnitude of $\Delta_{HTO}(1/H) \simeq |e|\hbar/(clm_e\varepsilon_{ov}) \simeq |e|\hbar/(cl'm_h\varepsilon_{ov})$, and smaller than that of SdH oscillations $\Delta_{ShH}^{e,h}(1/H) \simeq 2\pi|e|\hbar/(clS_{e,h})$. At low temperatures, the amplitude of the HTO of conductivity is small in the parameter $\sqrt{\hbar\omega_{e,h}/\varepsilon_{ov}}$ compared to the amplitude of the SdH oscillations σ^{ShH}, however for $T \geqslant \hbar\omega_{e,h}$ the amplitude of the most slowly decaying harmonics σ^{HTO} is exponentially large compared to σ^{ShH}.

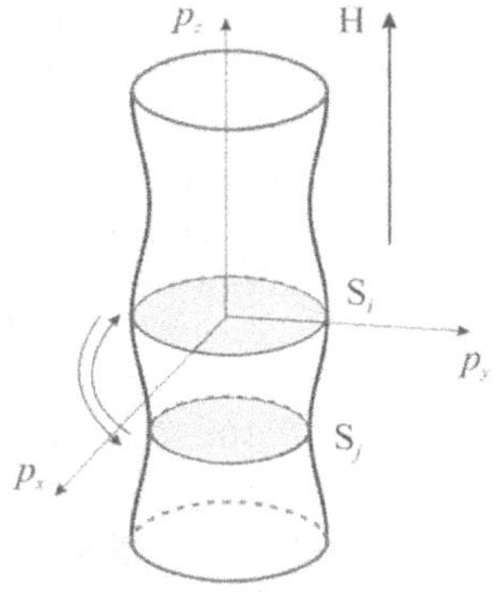

Fig. 8. Extreme sections of the Fermi surface and the electronic transitions responsible for high-temperature quantum oscillations in layered conductors.

The possibility of existence of the HTO in layered Q2D conductors has been shown in Ref. [61]. HTO arise due to the electron transition between the extreme FS sections with a similar area, as a result of scattering, see Fig. 8.

The periods of magnetic quantum oscillations are determined by a linear combination of the areas of the extreme sections. The HTO term of the transverse conductivity of a layered conductor with an arbitrary energy spectrum in a magnetic field directed along the normal to the layers in the case of elastic scattering by impurities and under condition (3.3), can be written as [61]

$$\sigma_{xx}^{HTO} = \sum_{i \neq j} \sigma_{ij}^{-}$$

$$\sigma_{ij}^{-} \propto (|S_i''||S_j''|)^{-1/2} \sum_{l,l'=1}^{\infty} \frac{(-1)^{l+l'}}{\sqrt{ll'}} R_T \left(\frac{2\pi^2 T}{\hbar \omega_{ll'}^{ij-}} \right) \cos\left(\frac{cS_{ll'}^{ij-}}{\hbar |e| H} - \frac{\pi}{4}(\alpha_i + \alpha_j) \right), \tag{4.2}$$

where

$$\omega_{ll'}^{ij-} = \frac{|e|H}{|m_{ll'}^{ij-}|c}, m_{ll'}^{ij-} = lm_i - l'm_j, S_{ll'}^{ij-} = lS_i - l'S_j, m_i = \frac{1}{2\pi} \frac{\partial S_i}{\partial \varepsilon}\bigg|_{\varepsilon = \varepsilon_F}$$

is the cyclotron mass of an electron in the i-th extreme FS section S_i, and, $\alpha_i = 1$ for minimum and $\alpha_i = -1$ for the maximum extreme sections. At low temperatures, the amplitude of the HTO approximately $\sqrt{\hbar\omega_H/(\eta\varepsilon_F)}$ times smaller than that of the SdH oscillations. As follows from Eq. (4.2), the HTO frequency $F^{HTO} \simeq c(S_{\max} - S_{\min})/(2\pi|e|\hbar) \simeq \eta m^* \varepsilon_F c/(2\pi|e|\hbar)$ and the argument of the function $R_T(x)$, describing the temperature decay, are determined by the difference between the two closely-located extreme sections, i.e., by the corrugation degree if the cylindrical FS. In contrast to semi-metals, the HTO period in layered structures is proportional to η^{-1} and significantly exceeds the period of the SdH oscillations.

The HTO have been found experimentally in β-(BEDT-TTF)$_2$IBr$_2$ [5, 6, 62], and a number of other layered organic conductors in a magnetic field of 10 T [63–66]. Fig. 9 shows the dependence of the resistance of β-(BEDT-TTF)$_2$IBr$_2$ in the direction of the lowest conductivity in a magnetic field, tilted by a small angle from the normal to the conductive layers [62]. At a temperature of 0.6 K, the SdH oscillations are observed for extreme orbits of the cylindrical FS with a frequency $F = c(S_{\max} + S_{\min})/(4\pi|e|\hbar)$. Just as in the case of the dHvA oscillations, due to the summation of the harmonics corresponding to the maximum and minimum cross sections of the

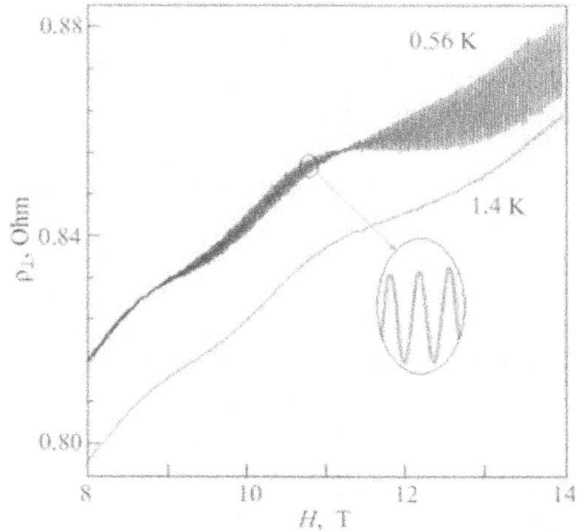

Fig. 9. Interlayer resistance of β-(BEDT-TTF)$_2$IBr$_2$ in a magnetic field tilted at an angle $\vartheta \approx 15°$ to the layer normal at different temperatures. The upper curve shows the SdH oscillations with a fundamental frequency $F \approx 3900$ T and a beat frequency $\Delta F \sim 20$ T. The lower curve shows the slow oscillations, which are weakly temperature dependent [62].

Fermi surface, there appear the beats with the frequency ΔF proportional to the difference between $S_{\max}$ and $S_{\min}$. The values of the FS two-dimensionality parameter η, determined from the ratio of the magnetoresistance beat frequency to the fundamental one, are in good agreement with the results obtained from experimental studies on the dHvA oscillations.

The dependence of the frequency $F^{HTO}(\vartheta)$ of the HTO of the interlayer resistance on the orientation of the magnetic field is similar to the angular dependence of the ΔF beat frequency for the SdH and dHvA oscillations [62]: $F^{HTO}(\vartheta)$ with changing ϑ, vanishing for the directions of **H** corresponding to the maxima of the quasi-classical part of the interlayer resistance. This behavior indicates that the HTO, similar to the beating of SdH and dHvA oscillations, arise due to the weak corrugation of the main FS cylinder. Experimental studies have shown that the relation $F^{HTO}(\vartheta) = 2\Delta F(\vartheta)$ holds in a wide range of angles with a sufficient accuracy. As follows from Reference [62], the amplitude of the HTO hardly changes with the temperature increase from 0.6 to 1.4 K, while the fundamental harmonic of the SdH oscillations disappears at 1.4 K.

5. Kinetic characteristics of layered conductors near the Lifshitz electronic topological transition

The density of states of elementary excitations in crystals $\nu(\varepsilon)$ has a root singularity for certain values of the energy $\varepsilon = \varepsilon_k$, and occurs with a change in the connectivity of constant-energy surfaces $\varepsilon(\mathbf{p}) = const$ [67]. At $\varepsilon = \varepsilon_k$ a new closed cavity of the surface nucleates, or narrow bridges on an open constant–energy surface break, or a bridge between the separated closed voids is formed. In metals and low-dimensional degenerate conductors with a high density of charge carriers, all the thermodynamic and kinetic characteristics at low temperatures are determined by the FS structure $\varepsilon(\mathbf{p}) = \varepsilon_F$ up to small corrections proportional to $(T/\varepsilon_F)^2$. Critical energy levels ε_k are significantly separated from the Fermi level ε_F, and the presence of root singularities $\nu(\varepsilon)$ for ε tending to ε_k in no way affects the electronic properties. In 1960, Lifshitz [20] suggested that the electronic topological transition in metals, accompanied by a change in the topological structure of the Fermi surface, is yet observable if it is possible to vary the chemical potential μ of electrons in a continuous manner, gradually bringing it closer to ε_k, for example, by applying a significantly high pressure or doping the conductor with impurity atoms of variable valence. For sufficiently close values of ε_F and ε_k, there occurs a substantial reconstruction of the topological structure of the Fermi surface. This electronic topological transition of the $2\frac{1}{2}$ kind, called the Lifshitz transition, is usually accompanied by the anomalous behavior of the thermodynamic and kinetic characteristics of the system of conduction electrons. The transition was soon discovered and actively investigated theoretically and experimentally in many metals and alloys in normal and superconducting states (more information can be found in the review Reference [68].

In spite of the assessment by Lifshitz [20] of the pressure necessary to observe a change in the connectivity of the Fermi surface, of the order of 10^4–10^5 kg/cm^2, which was difficult to access at that time, a considerably lower pressure was sufficient to observe the topological transition, and in some metals the anomalous behavior of the electron characteristics was observed in the absence of an external pressure. David Schoenberg at the satellite conference "Fermi Surface in Metals" to International Conference for Low Temperature Physics LT7 (Toronto, 1960) reported on the observation by his student Priestley of the magnetic susceptibility oscillations of magnesium in a magnetic field directed along the trigonal axis with a period corresponding to the cross section of the Fermi surface exceeding the dimension of the unit cell in the momentum space [69]. Because of the periodicity of the charge carrier dispersion law $\varepsilon(\mathbf{p})$, it follows that the conduction electrons responsible for this oscillation effect move from one unit cell to the other, tunneling through a thin interlayer between the individual cavities of the Fermi surface. Cohen and Fakicov used this phenomenon to interpret their measurements and called it magnetic breakdown [70]. A large number of experimental and theoretical studies were devoted to the study of the magnetization and the phenomena of charge transport in metals under conditions of possible magnetic breakdown in the sixties. The effect of magnetic breakdown on the oscillation dependence of the magnetization of metals on the magnitude of the strong magnetic field was investigated in detail by Pippard [71]. In the case of the appearance of magnetic breakdown open electron trajectories, the dependence of the magnetoresistance even of monocrystalline conductors on the magnetic field is close to linear [72, 73]. Reviews of the experimental studies of the magnetic breakdown phenomena in metals can be found, for example, in Refs. [74–76].

The Fermi surface in many layered conductors with a Q2D electron energy spectrum consists of multiple sheets, and a relatively small applied pressure is sufficient to observe the Lifshitz topological transition. Magnetic breakdown in quasi-two-dimensional conductors was first observed in the organic salt of terathiafulvalene κ-(BEDT-TTF)$_2$Cu (NCS)$_2$ [77, 78] in magnetic field of the order of 15 Tesla and soon in many other low-dimensional conductors, including heterostructures and various organic compounds [32, 34, 79–83]. For example, in Ref. [79], the authors reported the observation of angular magnetoresistance oscillations due to magnetic breakdown. The experiment has been conducted on an organic conductor κ-(BEDT-TTF)$_2$Cu(NCS)$_2$, using hydrostatic pressure and strong magnetic fields. According to the authors, the results have provided a convincing proof of the reliability of the semi-classical picture of the Pippard magnetic breakdown [71]. Such effects are likely to occur also in electron-doped cuprate superconducting materials [81–83]. The emergence of quasi-classical angular magnetoresistance oscillations and SdH oscillations in an electron-doped cuprate superconductor $Nd_{2-x}Sr_xCuO_4$ [81, 82], according to the authors, is a proof of the observation of magnetic breakdown in this material.

A statistical description of magnetic breakdown in conductors with the multi-sheet FS allows us to observe, on the background of complex movement of charge carriers, a rather deterministic nature of electron flow, similar to the diffusion fluxes of particles randomly wandering along Brownian trajectories. Let us consider a FS consisting of a corrugated cylinder and two quasiplanar sheets, weakly corrugated along the momentum projection p_z, while the p_x axis is orthogonal to the quasiplanar FS sheets (Fig. 10).

The current density in the τ-approximation for the collision integral has the form

$$J_i = \sigma_{ij} E_j = -\frac{2e^2 H}{c(2\pi\hbar)^3} \int d\varepsilon \frac{\partial f_0(\varepsilon)}{\partial \varepsilon} \int dp_H \int dt\, v_i(t, p_H) \psi(t, p_H) \equiv \langle v_i \psi \rangle, \tag{5.1}$$

For the variables in the momentum space, we take the integrals of motion of a charge in a magnetic field. Here, t is the time of motion

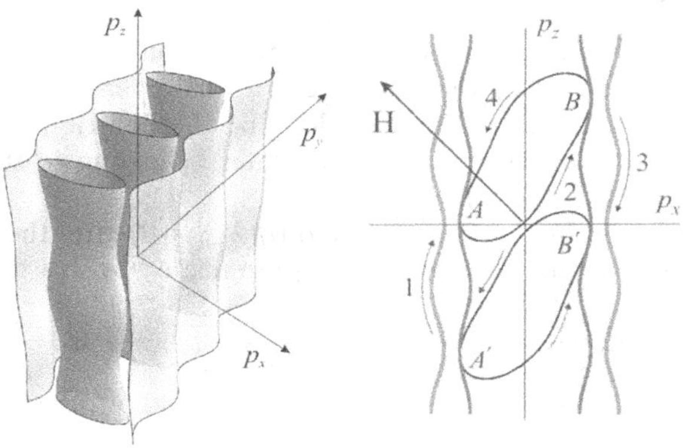

Fig. 10. Fermi surface and its projection on the $p_x p_z$ plane. The trajectories of the electrons belonging to two planar (arrows 1 and 3) and cylindrical (arrows 2 and 4) Fermi surface sheets in a magnetic field $\mathbf{H} = (H\cos\phi\sin\vartheta,\, H\sin\phi\sin\vartheta,\, H\cos\vartheta)$. A, B and A', B' are the regions of maximum approach of the Fermi surface sheets.

along the path $p_H = \mathbf{pH}/H = \text{const}$, $\varepsilon(\mathbf{p}) = \varepsilon_F$, $f_0(\varepsilon)$ is the equilibrium Fermi distribution function of the carriers. The function

$$\psi(t, p_H) = \int_{\lambda_1}^{t} e\mathbf{E}\mathbf{v}(t')\exp\frac{t' - t}{\tau}\,dt + \psi(\lambda_1, p_H)\exp\frac{\lambda_1 - t}{\tau}, \tag{5.2}$$

equals to the energy acquired by a conduction electron in the electric field $\mathbf{E}$, and the function

$$\psi(\lambda_1, p_H) = \int_{-\infty}^{\lambda_1} e\mathbf{E}\mathbf{v}(t)\exp\frac{t - \lambda_1}{\tau}\,dt \tag{5.3}$$

describes the entire history of the complex motion of an electron along the magnetic breakdown trajectories with the probability of magnetic breakdown w in the region A and the probability w' in the region B where individual FS cavities approach each other at times λ_1, λ_2, $\lambda_3, \ldots$. Here λ_1 is the closest to t moment of the transition of an electron from one FS sheet to another one, and $\lambda_j > \lambda_{j+1}$.

If there are several groups of carriers, each of them contributes to the current density

$$\langle \mathbf{v}\psi \rangle = \langle \mathbf{v}^{(1)}\psi_1 \rangle + \langle \mathbf{v}^{(2)}\psi_2 \rangle + \langle \mathbf{v}^{(3)}\psi_3 \rangle + \langle \mathbf{v}^{(4)}\psi_4 \rangle, \tag{5.4}$$

where $\langle \mathbf{v}^{(2)}\psi_2\rangle$ and $\langle \mathbf{v}^{(4)}\psi_4\rangle$ – are the contributions to the current due to the electrons, the states of which at a time t belong to different arcs 2 and 4 of the section of the weakly corrugated cylinder (see Fig. 9), and the remaining terms in (5.4) describe the contributions to the current due to the electrons with the states belonging to the flat FS sheets 1 and 3 (see Fig. 9).

The functions $\psi_i(\lambda_j - 0)$ before and $\psi_i(\lambda_{j+1} + 0)$ after the magnetic breakdown in the moment of time λ_{j+1} are related by

$$\psi_i(\lambda_j - 0) = A_i + \exp\left(\frac{\lambda_{j+1} - \lambda_j}{\tau}\right)\psi_i(\lambda_j + 0). \tag{5.5}$$

Here

$$A_i = \int_{\lambda_{j+1}}^{\lambda_j} e\mathbf{v}^{(i)}(t', p_H)\mathbf{E}\exp\left(\frac{t' - \lambda_j}{\tau}\right)dt' \tag{5.6}$$

is equal to the energy acquired by an electron in the electric field during the movement on the i-th sheet of FS between the two instants of the magnetic breakdown.

In the main approximation in the small parameter of two-dimensionality of the electron energy spectrum η, the functions A_{i} are the same for any value of λ_j. In the same approximation for the charge carriers beginning their random walk along the magnetic-breakdown trajectories from the first sheet of the Fermi surface, we have

$$\psi_1(\lambda_1 + 0) = (1 - w)(A_1 + \exp(-T_1/\tau)\psi_1(\lambda_2 + 0)) + w(A_2 + \exp(-T/\tau)\psi_2(\lambda_2 + 0)), \tag{5.7}$$

$$\psi_2(\lambda_2 + 0) = (1 - w)(A_4 + \exp(-T/\tau)\psi_4(\lambda_3 + 0)) + w(A_3 + \exp(-T_1/\tau)\psi_3(\lambda_3 + 0)), \tag{5.8}$$

$$\psi_3(\lambda_3 + 0) = (1 - w)(A_3 + \exp(-T_1/\tau)\psi_3(\lambda_4 + 0)) + w(A_4 + \exp(-T/\tau)\psi_4(\lambda_4 + 0)), \tag{5.9}$$

$$\psi_4(\lambda_4 + 0) = (1 - w)(A_2 + \exp(-T/\tau)\psi_2(\lambda_5 + 0)) + w(A_1 + \exp(-T_1/\tau)\psi_1(\lambda_5 + 0)), \tag{5.10}$$

$$\psi_1(\lambda_5 + 0) = (1 - w)(A_1 + \exp(-T_1/\tau)\psi_1(\lambda_6 + 0)) + w(A_2 + \exp(-T/\tau)\psi_1(\lambda_6 + 0)). \tag{5.11}$$

It is easy to see that Eq. (5.11) coincides with Eq. (5.7), but at an earlier point in time λ_5. Repeatedly applying the recurrence relations (5.7)–(5.10), we move to the distant past, and the sought functions in the right hand side of these equations gain exponentially small factors, while the terms containing A_i constitute a geometric progression, which can be easily summed up. As a result, we obtain

$$\psi_1(\lambda_1+0)=\frac{(1-w)A_1+wA_2}{1-h_1}+\sum_{n=0}^{\infty}h_1^n g\psi_2(\lambda_{n+2}+0), \qquad (5.12)$$

$$\psi_2(\lambda_1+0)=\frac{[A_1w(1+\gamma_1)+A_2(w(1-\gamma_1)+\gamma_1)](w'(1-\gamma_1)+\gamma_1)}{2ww'(\gamma+\gamma_1)+(w+w')\gamma_1(\gamma_1+2\gamma)+2\gamma\gamma_1^2}$$
$$+\frac{[A_3w'(1+\gamma_1)+A_4(w'(1-\gamma_1)+\gamma_1)](w+\gamma_1)(1+\gamma)}{2ww'(\gamma+\gamma_1)+(w+w')\gamma_1(\gamma_1+2\gamma)+2\gamma\gamma_1^2}. \qquad (5.13)$$

Here $h_1=(1-w)\exp(-T_1/\tau)$, $g=w\exp(-T/\tau)$, and T_1 is the period of motion of electrons on the quasi-planar FS sheet, T is the half-period of the motion along a closed cross section of the corrugated cylinder, $\gamma=\exp(T/\tau)-1$, and $\gamma_1=\exp(T_1/\tau)-1$. Functions $\psi_3(\lambda_j+0)$ and $\psi_4(\lambda_j+0)$ coincide with the functions of $\psi_1(\lambda_j+0)$ and $\psi_2(\lambda_j+0)$, when A_1 and A_3, A_2 and A_4, w and w' are pairwise interchanged, i.e $(A_1, A_2, w)\leftrightarrow(A_3, A_4, w')$. We have taken into account, that in the main approximation in the parameter η, the energy gained by electrons in an electric field between two events of possible magnetic breakdown is the same for any moment of magnetic breakdown λ_j. However in a magnetic field applied in yz-plane, there is $A_i(\lambda_j, \lambda_{j+1})=A_i(\lambda_1, \lambda_2)=A_i$ for any value of η. Using the formulas given above, we can calculate all the components of the electrical conductivity tensor σ_{ij} at any magnitude and orientation of a magnetic field.

For $\tan\vartheta\gg 1$, the sections of the FS by the plane $p_H=const$ are strongly elongated along the p_z axis, and the velocity v_z of an electron moving along the normal to the layers on this trajectory often changes the sign. The main contribution to its average value $\bar{v}_z$ during the period in a magnetic field is produced by small vicinities of the points of stationary phase, where

$$\frac{dp_z}{dt}=\frac{eH\sin\vartheta}{c}(v_x\sin\varphi-v_y\cos\varphi)=0. \qquad (5.14)$$

There are at least two such points on a closed section of a corrugated cylinder, and their contributions to the average velocity $\bar{v}_z$ can compensate each other at certain orientations of the magnetic field relative to the crystallographic axes of a single crystal sample. This leads to a sharp increase in the transverse resistance ρ_{zz}, which for $\eta \ll 1$ is asymptotically equal to $1/\sigma_{zz}$.

On the quasi-planar FS sheets, the points of stationary phase, satisfying the condition , are absent for $\varphi = \phi/2$ and usually appear only at a noticeable tilt of the magnetic field from the yz-plane. For $\varphi = 0$, i.e., for the magnetic field aligned in the xz-plane, there are two such points where p_x takes either the minimum value p_x^{min} or the maximum value p_x^{max}. From the period of these oscillations as a function of $\tan\vartheta$, it is possible to determine the magnitude of the corrugation of the planar FS sheet, $\delta p_x = p_{x1}^{max} - p_{x1}^{min} = p_{x3}^{max} - p_{x3}^{min}$ [19]. If an electron, during its free path, experiences a magnetic breakdown from one FS sheet to another at least once, then due to random walk of the charge carriers along the magnetic-breakdown trajectories combined frequencies of the angular oscillations of magnetoresistance appear [84].

Simple calculations allow us to obtain the following asymptotic expression for σ_{zz} [85] at $\tan\vartheta \gg 1$:

$$\begin{aligned}\sigma_{zz} = \frac{\sigma_0\eta^2}{\tan\vartheta}\{&\beta(1+\sin\alpha D_p) + 2\beta_1(1+\sin\alpha\delta p_x)\\ &+ \beta_2[2\cos\alpha(D_p+\delta p_x+\Delta_p) + \sin\alpha(D_p+2\delta p_x+2\Delta_p)\\ &\quad - \sin\alpha(D_p+\delta p_x)]\\ &+ \beta_3[\cos\alpha(\delta p_x+\Delta_p) - \sin\alpha\Delta_p + \sin\alpha(D_p+2\delta p_x+\Delta_p)\\ &\quad + \cos\alpha(D_p+\Delta_p)]\}. \end{aligned} \tag{5.15}$$

Here σ_0 is the electrical conductivity of a quasi-two-dimensional conductor along the layers in the absence of a magnetic field, D_p is the diameter of the cylinder along the p_x-axis, $\Delta_p = p_{x2}^{min} - p_{x1}^{max} = p_{x3}^{min} - p_{x2}^{max}$ is the minimum distance between the cylinder and flat sheets of the FS, $\alpha = (a/\hbar)\tan\vartheta$, the order of unity quantities β, β_1, β_2, and β_3 depend on the specific form of the electron energy spectrum. The asymptotic expression (5.15) for σ_{zz} is true when the corrugation along the p_x axis of the quasi-planar

FS sheet $\eta' = (\alpha\delta p_x)/\hbar$ is not small, and the condition

$$\eta' \tan\vartheta \gg 1, \quad (5.16)$$

is fulfilled. At the same time with increasing $\tan\vartheta$, the periods of the electron motion in a magnetic field, $T(\vartheta) = T(0)/\cos\vartheta$ and $T_1(\vartheta) = T_1(0)/\cos\vartheta$ increase, so that the observation of the angular magnetoresistance oscillations is possible only if

$$\gamma_0 \ll \cos\vartheta \ll \eta' \ll 1 \quad (5.17)$$

where $\gamma_0 = T(0)/\tau$.

The condition (5.17) is difficult to realize even in the case of very large mean free path of the charge carriers. Nevertheless, Kartsovnik et al. [80] have recently managed to observe the magnetic-breakdown oscillations of the interlayer magnetoresistance in an organic conductor α-(BEDT-TTF)$_2$KHg(SCN)$_4$ and α-(BEDT-TTF)$_2$TlHg(SCN)$_4$ with the quasi-one-dimensional electronic energy spectrum of one of the charge carrier groups. However, in the conductors with the highly corrugated FS in the plane of the layers, i.e., for η' of the order of unity, the contribution to the magnetic-breakdown oscillations comes from only a small fraction of the charge carriers from the vicinity of the points of the closest approach of the FS sheets. The probability of magnetic breakdown [86]

$$w(p_H) = \exp\left(-\frac{cS_p}{eH\hbar}\right) \quad (5.18)$$

decreases exponentially with increasing energy gap, which an electron has to overcome to get to the another part of the electron trajectory $\varepsilon(\mathbf{p}) = const$, $p_H = const$, along the line connecting the start point p_1 with the point of completion of the magnetic breakdown p_2. The value of the above-barrier area S_p in the case of small difference $(p_2 - p_1)$ coincides with the value $(p_2 - p_1)^2$ up to a numerical factor of the order of unity. Setting $(p_2 - p_1) \simeq \Delta_\varepsilon/v_F$, we obtain the formula for the probability of magnetic breakdown, calculated by Blount [87]

$$w = \exp\left(\frac{\kappa\Delta_\varepsilon^2}{\varepsilon_F\hbar\omega_H}\right), \quad (5.19)$$

where κ is the numerical coefficient, which depends on the specific form of the dispersion law of conduction electrons. In ordinary metals, the closest approach of individual sheets of the Fermi surface is small only on a small number of sections of the Fermi surface by the plane $p_H = const$, and on all other sections the value of $(p_2 - p_1)$ of the orders of the Fermi momentum p_F, and the probability of magnetic breakdown $w \simeq \exp(-\varepsilon_F/\hbar\omega_H)$ is negligible. However, the de Haas-van Alphen and Shubnikov-de Haas oscillations are formed by the fraction of electrons from a small neighborhood about $\delta p_H \leqslant p_F(\hbar\omega_H/\varepsilon_F)^{1/2}$ near the extremal section of the Fermi surface. For them, near the van Hove singularity, the probability of magnetic breakdown is of the order of unity. Therefore, it is not surprising that Priestley found in magnesium only the the de Haas-van Alphen magneto-breakdown oscillations (a more detailed information about measurement of the frequencies of magnetization oscillations is presented in Ref. [88]). In quasi-two-dimensional conductors, a much larger number of electrons form quantum oscillation effects, but the probability of magnetic breakdown $w = \exp(-\kappa_1\eta^2 r_H/a)$ of these charge carriers increases with decrease of the two-dimensionality parameter η. Here κ_1 is a factor of the order of unity, which depends on the dispersion law of charge carriers, r_H is the radius of curvature of its trajectory and as indicated above a is the distance between layers. At a sufficiently small approach of individual cavities of the Fermi surface in a magnetic field of the order of 30–50 Tesla, the probability of magnetic breakdown of electrons forming quantum oscillation effects is also of the order of unity. Because of the finite motion of the charge carriers along the magnetic breakdown trajectories, the resistance of the conductor increases linearly with the magnetic field in a sufficiently wide range of fields [89]. In two-dimensional layered structures, all charge carriers participate in the formation of various oscillatory effects in a strong magnetic field [90] with the same probability of magnetic breakdown.

The presence of the flat FS sheet leads to strong anisotropy of resistance in a strong magnetic field, even in the plane of the layers [91, 92]. If the probability of magnetic breakdown w is negligible, the component σ_{xx} of the resistance tensor is comparable with the conductivity σ_0 in the absence of a magnetic field. This occurs due to the presence, on a flat FS sheet, of open trajectories of the charge carriers drifting along the x-axis with an average speed $\bar{v}_x$. In this case the resistance in the plane of the layers increases with the square of the magnetic field and exhibits a deep minimum when the current flows along the x axis. When the planar FS sheets approach the slightly corrugated cylinder, the probability of magnetic breakdown evidently increases.

Starting from a planar sheet of the FS, an electron experiences a random walk through the magnetic-breakdown trajectories and weakens its ability to accelerate in an electric field E_x along the x-axis since the velocity v_x has an opposite sign on the opposite FS sheet. As a result, with increasing w the conductivity tensor component σ_{xx} decreases, leading to a substantial change in the dependence of the resistance along the layers on the magnitude of the strong magnetic field. For $w \geqslant \gamma$ the quadratic increase of the resistance in the plane of the layers with increasing the magnetic field changes to a linear dependence, which extends over a wide range of magnetic fields and reaches saturation only for $(1 - w) \leqslant \gamma$ [93]. The Hall field depends significantly on the magnetic breakdown probability, however its asymptote in the collisionless limit does not depend on τs for all values of w. For $w = 1$ the quasi-planar sheets of the FS touch the corrugated cylinders, and under further perturbation acting on the conductor, a rupture of the planar sheet along the line of contact occurs. As a result, the separated fragments of the planar FS sheet together with the halves of the corrugated cylinder eventually form a new corrugated cylinder, which is accompanied by the inversion of the charge carrier sign. This is not the only scenario of the Lifshitz topological transition, and the Hall effect study provides important information about the nature of changes in the topological structure of the electron energy spectrum at the Lifshitz phase transition.The discussion of a large number of experimental studies of magnetic breakdown phenomena in organic charge transfer complexes and an analysis of the most recent experiments can be found, for example, in J. Audouard and J.Y. Fortin article [94] and cited papers.

6. High-frequency resonances

The information on the energy spectrum and band structure of organic conductors can be obtained from studies of cyclotron resonance (CR) [95] and related phenomena associated with the resonant magnetic absorption of hf electromagnetic field. High-frequency resonances arise due to the periodic motion of conduction electrons in a magnetic field on the Fermi surface, in the case when their mean free time is large enough, and may be associated with the dynamics of Q2D [96–108] as well as Q1D [107–113] groups of charge carriers.

The first experiments on organic conductors using CR have been carried out on metals of the family (bis)ethylenedithio-tetritiafulvalena $(BEDT\text{-}TTF)_2MHg(SCN)_4$, ($M = K$, Tl, NH_4) [96–99]. CR has been first detected in the samples of $(BEDT\text{-}TTF)_2KHg(SCN)_4$ (Ref. [96]) in a magnetic field perpendicular to the conducting plane. For the frequencies of electromagnetic radiation $\nu = 316$–698 GHz, two reflectance minima have been observed with the resonant frequency linearly dependent on the magnetic field. The position of these features corresponds to the cyclotron masses $m_1 \sim 0.94\,m_e$ and $m_2 \sim 0.4\,m_e$, where m_e is the free electron mass [96]. Later, similar results have been obtained for other Q2D conductors, e.g., in a sample of $\alpha\text{-}(BEDT\text{-}TTF)_2NH_4Hg(SCN)_4$ at a frequency $\nu = 45$–65 GHz a resonance has been detected [97] with the cyclotron masses of $m \sim m_e$ and $m \sim 5\,m_e$. A family of organic metals $(BEDT\text{-}TTF)_2MHg(SCN)_4$ exhibits a characteristic antiferromagnetic ordering at helium temperatures. As a result, the structure of the resonant absorption of microwave radiation in these organic compounds is a superposition of peaks corresponding to the CR, against the background of which there appear more narrow lines with a 5–10 fold lower amplitude, presumably arising due to the electron paramagnetic (EPR) and antiferromagnetic (AFMR) resonances [100–102]. Under the conditions of these experiments, the radiation was almost completely absorbed by the crystal of the organic metal. Typical data on the power $P(H)$ absorbed by a sample of $\alpha\text{-}(BEDT\text{-}TTF)_2KHg(SCN)_4$ as a function of the magnetic field are shown in Fig. 11, [102].

Observable absorption maxima are grouped in series, depending on the behavior of the resonance frequency $\omega_r(H)$ with the magnetic field. The resonant frequency of the lines A and N increases with increasing H and does not extrapolate to the origin $\omega = 0$, $H = 0$. For the L line a decrease of $\omega_r(H)$ with increasing H has been observed. This behavior disagrees with the cases of CR and EPR, in which the extrapolation $H \rightarrow 0$ leads to $\omega \rightarrow 0$. According to the authors of Reference [102], lines A, N, and L are associated with the antiferromagnetic resonance. Line S is explained by ESR with a g-factor $g = 2.01$. Under the experimental conditions the movement of electrons into the sample occurred along the normal to the layers with the velocity η fold lower than the characteristic velocity of the electrons within the plane of the layers. For this reason, the diffusive mechanism of the emergence of EPR [114] is obviously ineffective. It is highly probable that the EPR

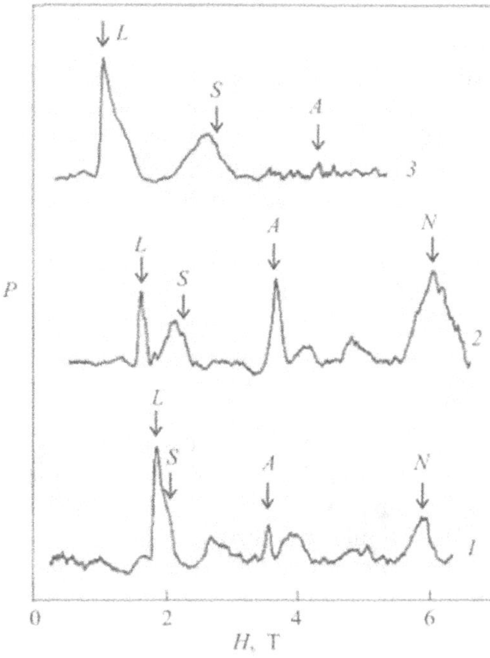

Fig. 11. Magnetic resonance in α-(BEDT-TTF)$_2$KHg(SCN)$_4$ in a magnetic field perpendicular to the sample surface at $T = 4.2$ K, [99]. Curves 1, 2, 3 correspond to the frequencies $\nu = 58$, 62, 78 GHz, respectively. The types of resonances are designated by letters.

absorption peaks in these studies are due to the excitation of spin collective modes.

Cyclotron resonance phenomenon, first observed in tetritiafulvalena salts, also takes place in Q2D organic conductors, the basic structural elements of which are other organic molecules. In Ref. [103], the absorption of microwave radiation has been measured in a single crystal of (BEDO-TTF)$_2$ReO$_4$(H$_2$O). This organic metal is structurally similar to the BEDT-TTF based compounds and has a pronounced resistance anisotropy: $\rho_a{:}\rho_b{:}\rho_c = 1{:}3{:}1000$ at a temperature of 300 K, where **a** and **b** are the crystallographic axis in the conducting plane, and the ***c*** axis is perpendicular to the **ab** plane. At low temperatures in the absence of a magnetic field (BEDO-TTF)$_2$ReO$_4$(H$_2$O) undergoes a transition into the superconducting state, however in the magnetic field of $H > 10\,\mathrm{kOe}$ the superconducting state is certainly destroyed and the conductor in question is a normal metal. An external magnetic field and the Poynting vector of the hf field were directed along the normal to the conductive plane in which the charge carriers move. The maximum absorption was observed at a frequency $\nu > 80\,\mathrm{GHz}$ at $T = 1.9\,\mathrm{K}$. The cyclotron mass of the charge carriers increased with the frequency of radiation from $0.8\,m_e$ ($\nu = 80\,\mathrm{GHz}$) to $0.95\,m_e$ ($\nu = 120\,\mathrm{GHz}$).

The effective masses of the charge carriers [100–102] found in cyclotron resonance experiments are significantly lower, approximately 2.5–3 fold, than the cyclotron mass determined from the temperature dependence of the amplitude of the SdH magnetoresistance oscillations. Even more significant is the difference of the mean free time of charge carriers in a static field and in an high-frequency field of the millimeter range. According to Ref. [103], the high-frequency mean free time is 30 times greater than that obtained in the low-frequency limit. This behavior might be explained by strong correlation effects in the electron Fermi liquid, resulting in a huge frequency-dependent renormalization of the effective mass.

The resonant absorption of the microwave field in organic conductors, associated with the quasi-periodic modulation of the electron velocity due to the corrugated Q2D or Q1D FS under high-frequency current in the direction of lower conductivity, is called the Periodic Orbit Resonance (POR) [107]. In Ref. [107] in the approximation of a uniform electric field, the interlayer conductivity has been calculated for a Q2D conductor with a FS in the form of a slightly corrugated cylinder with an elliptical dispersion of charge carriers in the conducting layers in a magnetic field directed along the normal to the layers. It has been shown that there are resonances of the interlayer conductivity at harmonics $n\omega_H$ of the main cyclotron frequency. The harmonic resonances up to the 7-th order have first been observed in a Q2D organic conductor $(\mathrm{BEDT\text{-}TTF})_2\mathrm{Br(DIA)}$ in a magnetic field perpendicular to the conductive plane [104]. Note that in plasma media with isotropic dispersion of charge carriers, the resonances of a higher order appear only when the spatial dispersion is taken into account.

The dependence of the interlayer conduction on the angle ϑ between the magnetic field and the normal to the layers is determined by the angular dependence of the cyclotron frequency

$\omega_H = (|e|H/m^* c)\cos\vartheta$ and the quasi-periodic modulation of the charge carrier velocity in the direction of lower conductivity. As the tilt angle of the magnetic field increases, the area of the closed section of the Fermi surface by the plane $p_H = const$ increases as $1/\cos\vartheta$ and, respectively, increases the rotation period of an electron on a cyclotron orbit. Starting from a certain value of ϑ, an electron cannot make a complete revolution along its orbit within the mean free time, and the condition of existence of the resonance $\omega_H \tau > 1$ ceases to be fulfilled. However, a resonance is possible when the magnetic field $H = (H\cos\phi, H\sin\phi, 0)$ is located in the plane of the conductive layers, i.e., when $\vartheta = \pi/2$. In this case, open trajectories appear in momentum space. For example, for the model dispersion relation corresponding to the free electron approximation for conductive-plane electrons and tight-binding band approximation for electrons belonging to adjacent layers, the dependence of the momentum projection p_z on time is determined by the equation of a physical pendulum, and the trajectory of an electron is described by the Jacobi elliptic functions. Resonance occurs when the oscillations frequency of the electron velocity coincides with the frequency of the electromagnetic wave ω. In the case of an elliptical dispersion law of electrons in the plane of the layers, the resonant value of the magnetic field is related with the maximum components v_{xm} and v_{ym} of the Fermi electron velocity along the long and short semiaxes of the elliptical cross section of the Fermi surface [106].

$$H_{res} = \frac{\hbar\omega c}{|e|a} \frac{1}{\sqrt{v_{xm}^2 \sin^2\phi + v_{ym}^2 \cos^2\phi}}, \tag{6.1}$$

where a is the distance between the layers. By performing the measurements at different orientations of the magnetic field, it is possible to determine the velocity field and the shape of the Fermi surface in the conducting plane [105,108].

The emergence of high-frequency resonance associated with the movement of electrons on open trajectories in a magnetic field in normal metals was first predicted theoretically in Ref. [115]. A similar effect in low-dimensional conductors, arising due to the dynamics of charge carriers on the Q1D part of the FS, was first observed experimentally [109] in an organic metal α-(BEDT-TTF)$_2$KHg(SCN)$_4$, the FS of which contains both sections and was dubbed the Fermi-surface traversal resonance (FTR). An electron moving along the quasi-one-dimensional FS section in the form of a slightly corrugated plane in a magnetic field perpendicular to the direction of the highest conductivity **a**, crosses the peaks and troughs of the FS sheet, and the velocity components in the directions **b** and **c** oscillate. As a result a resonance of the respective components of the high-frequency conductivity appears. In the case when the energy spectrum of charge carriers can be approximated by Eq. (2.7), the tensor of hf conductivity in the plane of best conductance **ab** in a magnetic field $\mathbf{H} = (0, H\sin\vartheta, H\cos\vartheta)$ in the absence of spacial dispersion takes the form:

$$\sigma_{xx} = \frac{ie^2\nu(\varepsilon_F)v_F^2}{\omega + i\tau^{-1}}, \quad \sigma_{xy} = \sigma_{yx} = 0,$$
$$\sigma_{yy} = ie^2\nu(\varepsilon_F)v_2^2\frac{\omega + i\tau^{-1}}{(\omega + i\tau^{-1})^2 - \Omega^2}, \tag{6.2}$$

If the mean free path length of the charge carriers is large enough so that $\Omega\tau \gg 1$, is provided

$$\omega = \Omega$$

the conductivity tensor component σ_{yy} has a sharp maximum, i.e., a high-frequency resonance emergence. Analogously to the Q2D case, we can write:

$$\Omega = \frac{|e|H}{cm_{1D}},$$

where m_{1D} is an analog of the effective cyclotron mass, which depends on the specific Q1D dispersion relation of the quasiparticle. In ordinary metals, the cyclotron mass can be characterized as a measure of how fast the electrons revolve in a closed orbit; accordingly, m_{1D} indicates how fast the electron propagates through the Brillouin zone, moving along an open trajectory on the Fermi surface. For an arbitrary electron energy spectrum, the magnitude of the magnetic field H_r at which the resonance occurs is given by

$$\frac{\omega}{H_r} = A\sin\psi, \tag{6.3}$$

where ψ is the angle between $\mathbf{H}$ and the **b**-axis and A is the proportionality factor.

Under the experimental conditions of Ref. [109], a single crystal sample was placed in a cavity of a size of $6 \times 3 \times 1.5$ mm with a quality factor of about 1000 at a temperature of 1.4 K and irradiated with a TE_{102} wave with a frequency $\nu \approx 70$ GHz. The induced current had components in the directions **b** and **c**. The depth of the skin layer was greater than the sample size, and spatial dispersion effects were not significant. The cavity transmission coefficient as a function of magnetic field for its different orientations exhibited minima corresponding to the maxima of conductivity.

High-frequency conductivity in the c direction of a Q1D conductor with the dispersion law , which is often used for the interpretation of experiments in a magnetic field $\mathbf{H} = (0, H\sin\vartheta, H\cos\vartheta)$ perpendicular to the conducting chain, is easy to calculate, using Eq. (2.11) for the components of the electron velocity and Eq. (2.12) for the tensor σ_{ij} in which the substitutions $\tau^{-1} \to \tau^{-1} - i\omega$ and $\omega_H \to \Omega$ should be made [46]

$$\sigma_{zz} = e^2\nu(\varepsilon_F)\tau \sum_{n,l} l^2 \frac{A_{nl}^2}{p_3^2} \frac{1}{1 + (\omega - (n - (p_2/p_3)l\tan\vartheta)\Omega)^2\tau^2}, \quad (6.4)$$

The conductivity (6.4) exhibits maxima when the frequency of the external field

$$\omega = \frac{|e|v_F H}{cp_2}\left(n\cos\vartheta - l\frac{p_2}{p_3}\sin\vartheta\right) \quad (6.5)$$

matches the frequencies of the Fourier harmonics of the electron velocity components. The parameters A_{nl}, n and m, determine the corrugation of the planar FS sheets. Analysis of the resistance minima in the direction of the lowest conductivity, corresponding to different harmonics of the velocity, allows us to determine the velocity of electrons on the Fermi surface and the parameters of the Q1D FS [108, 110–113].

7. Cyclotron resonance and angular impedance oscillations in Q2D organic conductors

In the experimental and theoretical studies of high frequency phenomena discussed in the previous section, the skin depth was comparable to or greater than the cyclotron radius, and the spatial dispersion of high-frequency electromagnetic field could be ignored. In this case, the collisionless absorption is due to the cyclotron absorption (for more details see the next paragraph). However, in organic metals, conditions of the anomalous skin effect can be realized, when spatial dispersion is essential.

Quasi-isotropic metals under the conditions of highly anomalous skin effect, when the penetration depth δ is much smaller than the Larmor radius r_0, the CR is most pronounced in a magnetic field parallel to the sample surface. In this case, all electrons repeatedly return to the skin layer, absorbing the energy of the electromagnetic field, and the surface impedance takes the minimum value when the wave frequency $\omega = n\omega_{H\,ext}$ $(n = 1, 2, 3, \ldots)$ is a multiple of the extreme value of the cyclotron frequency ω_H [95]. If the magnetic field is tilted away from the metal surface, most of the electrons visit the skin layer only once and then go into the bulk of the sample. Therefore, in the leading approximation in the small parameter δ/r_0, the resonance is absent, i.e., the oscillatory dependence of the impedance on the reciprocal of the magnetic field is manifested in the terms of higher order in δ/r_0 [116].

High-frequency resonant absorption in organic conductors in a strong magnetic field is characterized by a number of specific features. Collisionless absorption occurs due to electrons, the velocity $\mathbf{v}$ of which satisfies the equation:

$$\omega - n\omega_H - \langle \mathbf{kv} \rangle_\varphi = 0, \tag{7.1}$$

where $\mathbf{k}$ is the wave vector, the bracket $\langle \ldots \rangle_\varphi$ denotes averaging over the period $T = 2\pi/\omega_H$ of an electron moving in a magnetic field. The electron drift velocity $\mathbf{v}_D = \langle \mathbf{v} \rangle_\varphi$ in a strongly anisotropic conductive system can consist of the sum of a dc term $\bar{\mathbf{v}}_D$, independent on the momentum projection $p_H = (\mathbf{pH})/H$ and describing the motion along the open trajectories in momentum space, and an oscillating term $\tilde{\mathbf{v}}_D$, which depends on the angle ϑ between the magnetic field and the direction of the lowest conduction. The dc drift of conduction electrons leads to a Doppler shift of the resonance frequencies. The Landau damping of the ac electromagnetic field is determined by the term $\mathbf{k}\tilde{\mathbf{v}}_D$. Although $\tilde{v}_D$ is of the order of the characteristic velocity in the direction of the lowest conduction, the electron displacement for the period T can exceed the skin depth. In this case, $\mathbf{k}\tilde{\mathbf{v}}_D$ is of the order of ω_H order and the location of the Landau absorption regions is dependent on the orientation of $\mathbf{H}$, since $\tilde{\mathbf{v}}_D$ is an oscillating function of ϑ. As a result, angular oscillations of hf conductivity and surface impedance, which are caused by the angular dependence of the drift velocity of electrons, should appear [117]. If $\mathbf{k}\tilde{\mathbf{v}}_D \ll \omega_H$, the Landau damping can be neglected, and the angular dependence of the maxima of resonant absorption is determined by the angular dependence of frequency ω_H. For certain orientations of the magnetic field relative to the layers of the conductor, v_D is close to zero. For these directions $\mathbf{H}$, the Landau absorption is absent and even under the conditions of strongly anomalous skin effect, the cyclotron resonance of the same intensity as in a magnetic field parallel to the surface of the conductor occurs. In this case, the collisionless absorption of the electromagnetic field is caused by the absorption at the cyclotron frequency of an external electromagnetic field $\omega = n\omega_H$, equal to the cyclotron frequency and its harmonics.

As an example, let us consider Q2D metals of the family of Q2D tetrathiafulvalene salts. Their conductivity in the plane of the layers $\sigma_{\parallel}$ is comparable with the conductivity of ordinary metals; the ratio of rk to the conductivity across the layers $\sigma_{\perp}$ is typically of the order of 10^3–10^4. A simple estimate shows that for the frequencies of the order of 100 GHz, the condition $\tilde{v}_D T \geqslant \delta$ for the observation of the above phenomena can be easily realized.

Let us choose the XYZ coordinate system so that the z-axis is parallel to the direction of the lowest conductivity and the x-axis is perpendicular to the magnetic field $\mathbf{H} = (0, H\sin\vartheta, H\cos\vartheta)$ and the wave vector $\mathbf{k} = (0, k\sin\phi, k\cos\phi)$. In addition, we use another $x\xi\zeta$ coordinate system in which the n-axis is parallel to $\mathbf{k}$ and the conductor occupies the half-space $\xi > 0$.

In a tilted magnetic field, the resonance part of the current density occurs due to the electrons that do not collide with the sample boundary. Even for purely specular reflection, the projection of the momentum p_H is not preserved, and after the collision with the surface, the electron moves to a different orbit. If $\eta r_0 \simeq \delta$, then there are no electrons gliding over the surface of the conductor and thereby generating thereby a high-frequency conductivity resonance. For this reason, we can neglect the corrections to the conductivity tensor σ_{ij} due to the collisions of electrons with the surface and represent the current density in the form:

$$j_i(k) = \left(\sigma_{ik}(k) - \frac{\sigma_{i\xi}(k)\sigma_{\xi k}(k)}{\sigma_{\xi\xi}(k)}\right)\mathsf{E}_k(k),$$

$$\sigma_{ij}(k) = \frac{2|e|^3 H}{(2\pi\hbar)^3 c}\int \frac{\omega_H^{-2}\,dp_H}{1 - e^{2\pi i(\tilde{\omega} - \langle \mathbf{kv}\rangle_\varphi)/\omega_H}}$$

$$\times \int_0^{2\pi} d\varphi v_i(\varphi) \int_0^{2\pi} d\varphi_1 v_j(\varphi - \varphi_1) e^{i(\tilde{\omega}/\omega_H)\varphi_1 - iR(\varphi,\varphi_1)}. \qquad (7.2)$$

Here $\{i,k\} = \{x,\zeta\}$, $\tilde{\omega} = \omega + i\tau^{-1}$, $\varphi = \omega_H t$ is the phase of the electron velocity, and

$$R(\varphi,\varphi_1) = \omega_H^{-1} \int_{\varphi-\varphi_1}^{\varphi} d\varphi' \mathbf{kv}(\varphi').$$

To obtain a simple analytical expression for conductivity and plot the resonance dependence of the surface impedance on the magnitude and direction of an external magnetic field, let us apply the model electron energy spectrum which corresponding to the free electron approximation for in-plane electrons and tight-binding band approximation for electrons belonging to adjacent layers

$$\varepsilon(\mathbf{p}) = \frac{p_x^2 + p_y^2}{2m} - \varepsilon_0 \cos\frac{p_z}{p_0} \qquad (7.3)$$

here $\varepsilon_0 \simeq \eta\varepsilon_F$, $m = const$ is the effective mass in the plane of the layers. In the case of an arbitrary Q2D cylindrical FS, the shape of the resonance lines changes slightly, but the qualitative properties of the resonant behavior of the impedance remain.

For $\eta \tan\vartheta \ll 1$, the dependence on p_H of the velocity components v_x and v_y in the conductive plane

$$v_x^{(0)}(t) = -v_\perp \sin\omega_H(\beta)t, \quad v_y^{(0)}(t) = v_\perp \cos\omega_H(\beta)t, \qquad (7.4)$$

and the cyclotron frequency

$$\omega_H(\beta) = \omega_H^{(0)}(1 + \eta\tan\vartheta\, \mathrm{J}_1(\alpha)\cos\beta) \qquad (7.5)$$

appears in the first approximation in $\eta\tan\vartheta$. Here $\omega_H^{(0)} = (|e|H/mc)\cos\vartheta$, $\alpha = (mv_F/p_0)\tan\vartheta$, $\beta = p_H/p_0\cos\vartheta$, $\mathrm{J}_n(\alpha)$ are

Bessel functions; the initial phase chosen such that $v_x(0) = 0$, $v_\perp = v_F(1 + O(\eta))$ is the amplitude of the first harmonic. The velocity along the normal to the layers v_z is much less than the Fermi velocity v_F of the electrons

$$v_z(t) = \eta v_F \sin(\beta - \alpha \cos \omega_B(\beta)t). \tag{7.6}$$

The drift velocity of an electron in the first approximation in η.

$$\mathbf{v}_D \equiv \langle \mathbf{v} \rangle_\varphi = \frac{1}{2\pi}\int_0^{2\pi} d\varphi\, \mathbf{v}(\varphi)$$

$$= (\mathbf{e}_y \tan\vartheta + \mathbf{e}_z)\langle v_z \rangle_\varphi = \eta v_F \mathrm{J}_0(\alpha)(\mathbf{e}_y \tan\vartheta + \mathbf{e}_z)\sin\beta \tag{7.7}$$

contain the Bessel functions of the argument $(mv_F/p_0)\tan\vartheta$, which oscillates with varying ϑ.

Under conditions of a strongly anomalous skin effect, when $kv_F \sin\phi \gg \omega_H$, ω, the integrals with respect to φ and φ_1 in Eq. (7.2) can be calculated by the stationary phase method [118]. Since $kv_z \simeq \eta k v_F$, the stationary points are determined by the equations

$$v_\xi^{(0)}(\varphi) = v_y^{(0)}(\varphi)\sin\varphi = 0, \quad v_\xi^{(0)}(\varphi - \varphi_1) = 0.$$

It is easy to see that at $kr_0 \gg \omega\tau$, the maximal component of the tensor σ_{ij} is σ_{xx}, which is proportional to $(kr_0)^{-1}$, $r_0 = v_F/\omega_H$. The power series expansion of the component σ_{yi}, $i = x, y, z$ starts with higher-order terms in $(kr_0)^{-1}$ and $|\sigma_{xy}|^2$ is small compared to $|\sigma_{xx}\sigma_{yy}|$. The components σ_{zi}, $(i = x, y)$ are proportional to η, and the component $\sigma_{zz} \sim \eta^2$.

In the leading approximation in the small parameters $(kr_0)^{-1}$ and η, the maximum component of the conductivity tensor has the form

$$\sigma_{xx}(k) = \frac{i\omega_p^2}{4\pi^2\omega_H k_y r_0}\int_{-\pi}^{\pi} d\beta \frac{\cos\frac{\pi}{\omega_H}(\tilde{\omega} - \langle \mathbf{kv} \rangle_\varphi) - \sin R_0}{\sin\frac{\pi}{\omega_H}(\tilde{\omega} - \langle \mathbf{kv} \rangle_\varphi)} \tag{7.8}$$

where $\omega_p = \sqrt{4\pi n_0 e^2/m}$ is the plasma frequency, n_0 is the electron density,

$$R_0 = \omega_H^{-1}\int_{-\pi/2}^{\pi/2} d\phi\, \mathbf{k}\tilde{\mathbf{v}}(\varphi) \approx 2k_y v_F/\omega_H = kd_0,$$

$d_0 = 2r_0 \sin\varphi$ is the displacement of an electron along the ξ axis during the half period π/ω_H, $\tilde{\mathbf{v}} = \mathbf{v} - \langle\mathbf{v}\rangle_\varphi$, and $\beta = p_H/(p_0\cos\vartheta)$. In the first approximation in η the average $\langle\mathbf{kv}\rangle_\varphi$ is

$$\langle\mathbf{kv}\rangle_\varphi = \eta k v_F \mathrm{J}_0(\alpha)(\sin\phi\tan\vartheta + \cos\phi)\sin\beta, \tag{7.9}$$

where $\alpha = (mv_F/p_0)\tan\vartheta$, and $\mathrm{J}_0(\alpha)$ is the Bessel function.

In the collisionless limit $\tau^{-1} \to 0$, Eq. (7.8) can be written as

$$\sigma_{xx} = \frac{\omega_p^2}{\pi k d_0 \omega_H}\left\{\sum_{n'} \frac{1-(-1)^{n'}\sin kd_0}{\sqrt{(\eta k_1 v_F \mathrm{J}_0(\alpha))^2 - (-n'\omega_H)^2}} + i\sum_{n''}\frac{\operatorname{sign}(\omega - n''\omega_H)(1-(-1)^{n''}\sin kd_0)}{\sqrt{(\omega - n''\omega_H)^2 - (\eta k_1 v_F \mathrm{J}_0(\alpha))^2}}\right\} \tag{7.10}$$

The summation in Eq. (7.10) is carried out over n' and n'' such, that $(\omega - n'\omega_H)^2 - \langle\mathbf{kv}\rangle_\phi^2 < 0$ and $(\omega - n''\omega_H)^2 - \langle\mathbf{kv}\rangle_\phi^2$, respectively, and $k_1 = k(\sin\phi\tan\vartheta + \cos\phi)$.

Conductivity in Eqs. (7.8), and (7.10) is an oscillatory function of the angle ϑ since $\langle\mathbf{v}\rangle_\varphi$ u is proportional to $\mathrm{J}_0(\alpha)$. Sharp maxima occur for $\vartheta = \vartheta_m$, for which the condition (7.1) holds. In the vicinity of these angles, the surface impedance

$$Z_{xx} = -\frac{8i\omega}{c^2}\int_0^\infty \frac{dk}{k^2 - 4\pi i\omega c^{-2}\sigma_{xx}(k)} \equiv RiX, \tag{7.11}$$

as a function of ϑ, takes the minimum values. The angular dependence of the real and imaginary parts Z_{xx} at $\omega = \omega_0 = |e|H/mc$ is shown in Fig. 12.

The physical nature of the quasi-periodic dependence of the transport coefficients on the angle ϑ manifests itself in the emergence of strong collisionless absorption for certain directions of **H**. An essential criterion for the observation of this effect as well as for other electronic transport phenomena in magnetic field is $\omega_H\tau > 1$. Therefore, the oscillation amplitude of the impedance decreases with increasing ϑ due to lowering the cyclotron frequency, which is proportional to $\cos\vartheta$.

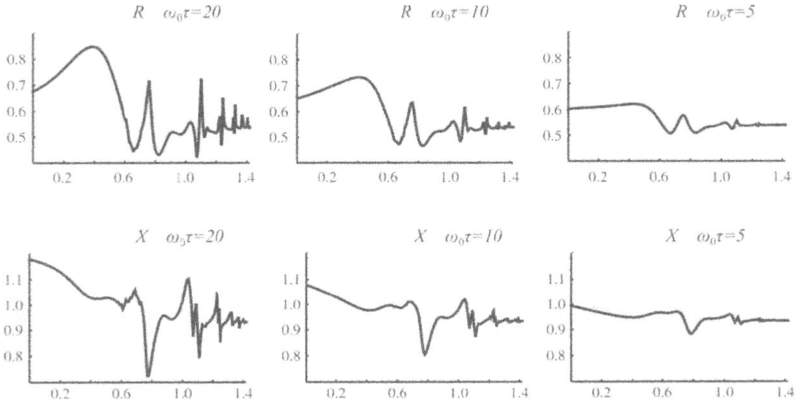

Fig. 12. Dependence of the R/Z_0 and X/Z_0 on ϑ, where $Z_0 = 8\omega/c^2k_0$, $k_0 = (2\omega_p^2\omega_0/v_Fc^2)^{1/3}$, when $\omega = \omega_0 = |e|H/mc$, $mv_F/p_0 = 3$, $k_0v_F/\omega_0 = 50$, $\eta = 0.02$, $\phi = \pi/4$, $\omega_0\tau = 20$, 10, 5.

A characteristic feature of the high-frequency resonances in Q2D metals is the weak difference of the electron motion periods at different sections of the FS by the plane $p_H = const$. Unlike conventional metals, in which $kv_D \gg \omega_H$ for almost all points on the FS, in quasi-2D conductors in a tilted magnetic field, the oscillating dependence of the impedance on H^{-1} manifests itself in the leading approximation in δ/r_0. For $kv_D \simeq \eta kv_F \simeq \omega_H$, under resonance conditions the conductivity exhibits a root singularity, and the amplitude of the oscillations of the impedance increases with decreasing anisotropy of the Fermi surface. The latter occurs due to the decrease of Landau damping. In conductors with a small anisotropy parameter $\eta kv_F \ll \omega_H$ or for the directions of $\mathbf{H}_0$, for which v_D is close to zero, the Landau absorption is absent and even under the conditions of the anomalous skin effect, the cyclotron resonance is of the same intensity as in the magnetic field parallel to the sample surface. In this case, the collisionless absorption of the electromagnetic field is caused by the cyclotron absorption at the frequency of the external electromagnetic field $\omega = n\omega_H$ equal to the cyclotron frequency and its harmonics. The real and imaginary parts of the impedance Z_{xx} as functions of ω/ω_H at $\omega\tau = 20$ for different values of the anisotropy parameter η are shown in Fig. 13, [119].

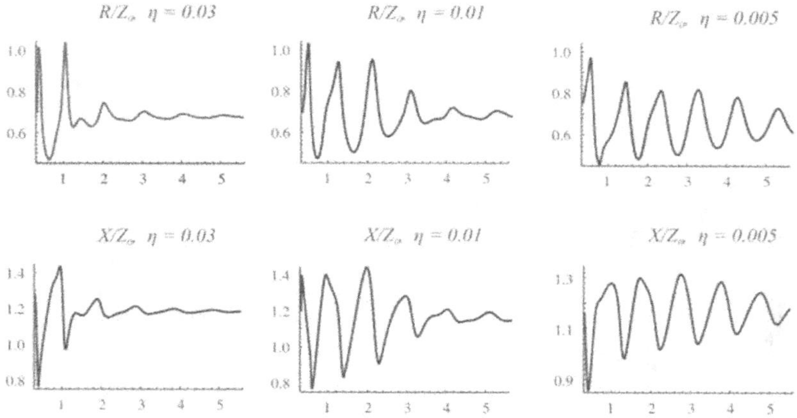

Fig. 13. Dependence of the R/Z_0 and X/Z_0 on ω/ω_H, where $Z_0 = 8\omega/c^2k_0$ $k_0 = (2\omega_p^2\omega_0/v_Fc^2)^{1/3}$, for $\omega\tau = 20$, $k_0v_F/\omega = 50$, $\vartheta = \pi/6$, $\phi = \pi/4$, $mv_F/p_0 = 2$, $\eta = 0.03$, 0.01, 0.005.

For those directions of **H**, when α is equal to one of the roots of the Bessel function $J_0(\alpha)$, or when the condition $\eta k v_F \ll \omega_H$ is fulfilled, Eq. (7.8) is transformed into

$$\sigma_{xx} = \frac{i\omega_p^2}{\pi k d_0 \omega_H}\left(\cot(\pi\tilde{\omega}/\omega_H) - \frac{\sin kd_0}{\sin(\pi\tilde{\omega}/\omega_H)}\right), \tag{7.12}$$

and the expressions for the electric field in a conductor

$$E_x(\xi) = -\frac{2E'_x(0)}{\pi}\int_0^\infty \frac{dk \cos k\xi}{k^2 - 4\pi i\omega c^{-2}\sigma_{xx}(k)} \tag{7.13}$$

and the surface impedance (7.11) can be significantly simplified. Here, $E_i'(0)$ is the derivative of the electric field on the conductor surface. Since the real part of the components $\sigma_{xx}(k)$ of the conductivity tensor is positive, we can write

$$\frac{k}{k^3 - 4\pi i k\omega c^{-2}\sigma_{xx}} = ik\int_0^\infty dt e^{-i(k^3 - 4\pi i k\omega c^{-2}\sigma_{xx})t}. \tag{7.14}$$

Expanding the exponential into the Fourier series in kd_0 and integrating the result with respect to t, we obtain

$$E_x(\xi) = -\frac{2iE'_x(0)}{\pi}\int_0^\infty \frac{dkk}{\sqrt{a^2+b^2}}$$
$$\times\left\{\cos k\xi + \sum_{l=1}^{\infty} i^l b^l \frac{\left(\begin{array}{c}\cos\left[k(\xi+ld_0)-\frac{\pi l}{2}\right]\\ +\cos\left[k(\xi-ld_0)-\frac{\pi l}{2}\right]\end{array}\right)}{(\sqrt{a^2+b^2}+a)^l}\right\}. \quad (7.15)$$

where

$$a = ik^3 + i\frac{k_0^3\,\omega}{\omega_0 \sin\phi}\cot\frac{\pi\tilde{\omega}}{\omega_H},\quad b = \frac{k_0^3\,\omega}{\omega_0\sin\phi}\sin^{-1}\frac{\pi\tilde{\omega}}{\omega_H},$$
$$k_0 = (2\omega_p^2\omega_0/v_F c^2)^{1/3},\quad \omega_0 = |e|H_0/mc.$$

Since k_0 is of the order of δ^{-1}, the integrand in Eq. (7.15) is a rapidly oscillating function, and the electric field $E_x(\xi)$ takes the extreme values at distances close to ld_0 from the sample surface.

In the collisionless limit the conductivity (7.12) becomes imaginary, and weakly damped waves with frequencies $\omega = n\omega_H + \Delta\omega(k)$, $0 < |\Delta\omega(k)| < \omega_H$ can propagate even under conditions of strong spatial dispersion [120].

For $\vartheta = \vartheta_i$ such that $\mathbf{v}_D \approx 0$, the impedance

$$Z_{xx} = \frac{8}{3ic}\left(\frac{\omega^2 v_F \sin\varphi}{2\omega_p^2 c}\right)^{1/3}\int_0^\infty \frac{dt t^{-1/3}}{\sqrt{t^2 - 1 + 2t\cot\frac{\pi\tilde{\omega}}{\omega_H}}} \quad (7.16)$$

exhibits an oscillatory dependence on the reciprocal of the magnetic field, which is typical for conductors placed in a magnetic field parallel to the sample surface. In the vicinity of resonances when

$$|\omega - n\omega_H| \ll \omega_H,\quad \left|\cot\frac{\pi\tilde{\omega}}{\omega_H}\right| \gg 1,$$

the asymptotic representation (7.16) has the form

$$Z_{xx} \approx -\frac{8i}{3c}\left(\frac{\omega^2 v_F\sin\phi}{2\omega_p^2 c}\right)^{1/3} \mathrm{B}\left(\frac{1}{6},\frac{2}{3}\right)\tan^{1/3}\frac{\pi\tilde{\omega}}{\omega_H}, \quad (7.17)$$

where $\mathrm{B}(x,y)$ is the Euler beta function.

Under strict resonance conditions $\omega = n\omega_H$ and $\vartheta = \vartheta_i$, the electric field can be written as

$$E_x(\xi) = \sum_l E_l(\xi), \tag{7.18}$$

where the functions $E_l(\xi)$ are significantly different from zero only in a small neighborhood $\xi = ld_0$,

$$E_l(ld_0) = (-1)^{ln+1} \frac{e^{i\pi/6} \cos\left(\frac{\pi l}{2}\right) \tan h^{1/3}\gamma}{2^{l-1} 3\pi n^{1/3}\, \kappa_0 \cosh^l \gamma} E_x'(0)$$

$$\mathrm{B}\left(l + \frac{1}{3}, \frac{2}{3}\right) {}_2\mathrm{F}_1\left(\frac{l}{2} + \frac{1}{6}, \frac{l}{2} + \frac{2}{3}, l + 1, \frac{1}{\cosh^2 \gamma}\right) \tag{7.19}$$

where $\kappa_0 = (4\omega_p^2/d_0c^2)^{1/3}$, $\gamma = \pi/(\omega_H\tau)$ and ${}_2\mathrm{F}_1(\alpha, \beta, \gamma, z)$ is the Gauss hypergeometric function.

The structure of the electric field in the bulk of the sample is similar to that predicted by Azbel [121] for quasi-isotropic metals placed in a magnetic field parallel to the sample surface. For even l there are single peaks, the signs of which alternate as the $\cos(\pi l/2)$. For odd l there are two maxima in the vicinity of $\xi = ld_0$, which differ in the sign, and the electric field is antisymmetric with respect to $\xi = ld_0$. The difference in the field distribution in Q2D conductors is that a decrease in the intensity of the peaks of electromagnetic field with l is much slower. Nearby peaks are of the same order of magnitude. This is due to the fact that the surges [121] are caused by electrons belonging the extreme sections of the FS, whereas in Q2D conductors, essentially all the electrons with the Fermi energy are involved in their formation. As follows from Eq. (7.19) the ratio of the amplitudes of adjacent surges is of the order of

$$\left|\frac{E((l+1)d_0)}{E(ld_0)}\right| \simeq \exp\left(-\frac{\pi}{\omega_H\tau}\right) \tag{7.20}$$

The amplitude of the peaks decreases with increasing ϑ_i since γ is proportional to $1/\cos\vartheta$.

If the direction of the lowest conductivity coincides with the inner normal to the surface of the conductor, i.e., $\phi = 0$, the integrand in Eq. (7.2) does not have a rapidly oscillating phase since $\mathbf{kv} \simeq \eta k v_F$ is proportional to η and we cannot obtain simple asymptotes of the transport coefficients. However, also in this case, the physical mechanism discussed above results in resonant oscillations of the transport coefficients with changing the magnetic field orientation if the condition $\eta k r_0 \simeq 1$ (or $\eta r_0 \simeq \delta$) is satisfied.

8. Cyclotron waves in Q2D conductors

In metals at low temperatures different electromagnetic collective modes can exist, a large part of which have analogues in the gas plasma. Most of these excitations are strongly damped, and only in certain frequency ranges and for specific parameters of the solid-state plasma, weakly damped plasma waves can exist. In the absence of a magnetic field, electromagnetic waves with frequencies of lower the plasma frequency ω_p cannot propagate in the plasma media; they decay or undergo total reflection. The absorption of the waves is caused by electron collisions and collisionless Landau damping, which is the resonant absorption of electromagnetic field by the charge carriers the velocity of which along the wave vector coincides with the phase velocity of the wave. The latter absorption mechanism is essential for high-frequency modes $\omega\tau \gg 1$. For $\omega > \omega_p$, the bias current exceeds the conduction current, the dielectric permittivity in the collisionless limit $\tau \to \infty$ is positive, and the plasma is transparent to electromagnetic waves.

A magnetic field affects the dynamics of electrons and changes the electromagnetic properties of the plasma medium. At low temperatures in conductors placed in a magnetic field, the propagation of waves with frequencies of much lower ω_p and the attenuation length equal to the mean free path of the charge carriers is possible, provided that an electron during the mean free time completes at least a few turns on the cyclotron orbit [122, 123].

In the presence of a dc magnetic field, there appears another mechanism of collisionless absorption – cyclotron damping that occurs when the frequency of the electromagnetic field is equal to cyclotron frequency ω_H or its harmonics $n\omega_H$. Electrons moving spirally in phase with the wave are accelerated by an electric field in the plane perpendicular to **H** and absorb the energy of the electromagnetic field. At the same time the resonant particles moving along **H** still cause Landau damping. Sharp peaks of high-frequency conductivity corresponding to intense collisionless absorption appear under the conditions of Eq. (7.1).

As a rule, weakly damped waves are associated with high-frequency resonances. The electromagnetic energy absorbed by a conductor under resonance condition can propagate in the form of collective modes. The type of excitations characteristic for both solid and gas plasmas are the so-called cyclotron waves. This term commonly denotes collective modes the frequency of which is close to the rotation frequency of electrons in a magnetic field and its harmonics $\omega = n\omega_H + \Delta\omega$, $n = 1, 2, \ldots$, where $|\Delta\omega| > \tau^{-1}$. Their physical nature can be explained by considering electrons in a magnetic field as oscillators with the eigenfrequency ω_H, which act as sources of an electromagnetic field. The harmonics of the cyclotron frequency are excited since the motion of an electron in its orbit in a dc magnetic field is perturbed by a non-uniform high-frequency self-consistent field. In ordinary metals and gas plasma, cyclotron waves propagate predominantly perpendicular to the external magnetic field under condition of a non-local coupling of the current density and the electric field. In the case $\mathbf{k} \perp \mathbf{H}$, the region of collisionless damping in the plane (ω, k) is transformed into a series of lines, and the resonance absorption occurs only when the condition $\omega = n\omega_H$ is strictly satisfied.

The FS topology has a significant effect on the characteristics of the collective modes since the electric current excited by the wave has a self-consistent effects on the wave process. The propagation of the collective modes in layered Q2D conductors has a number of features associated with the energy spectrum of the charge carriers. For certain orientations of a magnetic field relative to the layers of the conductor, the projection of the electron velocity on the direction of **H**, averaged over the period of the cyclotron orbit, is a negligible quantity. For these directions of **H**, collisionless absorption is absent and the propagation of weakly damped waves is possible even under conditions of strong spatial dispersion [120] for an arbitrary orientation of the wave vector with respect to **H**. Thus, in strongly anisotropic organic conductors, there are certain transparency directions for propagation of the collective modes under conditions of strong spatial dispersion. This feature is common to all elementary excitations of the Bose type in the conducting systems under consideration.

In this paragraph we consider the cyclotron waves in layered conductors with a quasi-two-electron energy spectrum [120] for an arbitrary orientation of the magnetic field and the wave vector with respect to the layers. The drift of an electron along the magnetic field during the rotation period at the cyclotron orbit is assumed to be of the order of magnitude of the wavelength of the electromagnetic field in the conductor. Under these conditions, the special dispersion in the direction of the lowest conductivity should be considered.

Assuming that the space-time dependence of all the variables has the form $\exp(i\mathbf{kr} - i\omega t)$, it is straightforward to obtain the dispersion relation determining the frequencies of the eigenmodes $\omega(\mathbf{k})$ of the electromagnetic field using Maxwell's equations,

$$\det\left[k^2\delta_{ij} - k_i k_j - \frac{\omega^2}{c^2}\varepsilon_{ij}(\omega, \mathbf{k})\right] = 0, \tag{8.1}$$

The real parts of the roots of Eq. (8.1) determine the spectrum of collective modes, and the imaginary parts describe the damping rate. Here, $\varepsilon_{ij}(\omega, \mathbf{k}) = \delta_{ij} + (4\pi i/\omega)\sigma_{ij}(\omega, \mathbf{k})$ is the permittivity tensor and δ_{ij} is the Kronecker symbol. For the frequencies ω much smaller than δ_{ij}, the first term in the expression for ε_{ij} can be neglected.

If the wave vector $\mathbf{k} = (k\sin\phi, 0, k\cos\phi)$ lies in the xz plane, then by calculating the determinant, we obtain

$$A\left(\frac{kc}{\omega}\right)^4 + B\left(\frac{kc}{\omega}\right)^2 + C = 0, \tag{8.2}$$

where

$$A = \varepsilon_{xx}\sin^2\phi + 2\varepsilon_{xz}\sin\phi\cos\phi + \varepsilon_{zz}\cos^2\phi,$$

$$B = -\varepsilon_{xx}\varepsilon_{zz} - (\varepsilon_{yy}\varepsilon_{zz} + \varepsilon_{yz}^2)\cos^2\phi - (\varepsilon_{xx}\varepsilon_{yy} + \varepsilon_{xy}^2)\sin^2\phi + \varepsilon_{xz}^2$$
$$+ 2(\varepsilon_{xy}\varepsilon_{yz} - \varepsilon_{xz}\varepsilon_{yy})\sin\phi\cos\phi$$

$$C = \det[\varepsilon_{ij}] = \varepsilon_{zz}(\varepsilon_{xx}\varepsilon_{yy} + \varepsilon_{xy}^2) + \varepsilon_{xx}\varepsilon_{xy}^2 + 2\varepsilon_{xy}\varepsilon_{xz}\varepsilon_{yz} - \varepsilon_{yy}\varepsilon_{xz}^2$$

In the most general form, the Q2D energy spectrum of charge carriers in the one electron approximation is determined by Eq. (2.1). In the case of strong spatial dispersion, $kr_0 \gg 1$, $\eta k r_0 \simeq 1$, where $r_0 = v_F/\omega_H$ is the Larmor radius of a conduction electron (as in the previous sections, here $\omega_H = |e|H/m^*c$, where m^* – is the cyclotron effective mass of charge carriers). The integrals over t and t_1 in Eq. (7.2) are calculated by the method of stationary phase [118].

Furthermore if $\omega \simeq \omega_H$, the stationary points are determined from the equations $v_x(\varphi) = 0$, $v_x(\varphi - \varphi_1) = 0$, where $\varphi = \omega_H t$, $\varphi_1 = \omega_H t_1$. In the most general case, the asymptote of the conductivity tensor has the form

$$\sigma_{ij}(\omega, \mathbf{k}) = \frac{e^2}{2\pi^2\hbar^3}\int dp_H \frac{m^*\omega_H^{-1}}{1 - \exp\left(2\pi i \dfrac{\tilde{\omega} - \langle \mathbf{kv}\rangle_\varphi}{\omega_H}\right)}$$

$$\times \sum_\alpha \kappa(\boldsymbol{\varphi}^{(\alpha)}) \frac{\exp\left[i\dfrac{\tilde{\omega}}{\omega_H}\varphi_1^{(\alpha)} - iR(\varphi^{(\alpha)}, \varphi_1^{(\alpha)}) + i\dfrac{\pi}{4}s\right] \times v_i(\varphi^{(\alpha)})v_j(\varphi^{(\alpha)} - \varphi_1^{(\alpha)})}{\sqrt{|\det(\mathrm{R}''_{\varphi\varphi_1}(\varphi^{(\alpha)}, \varphi_1^{(\alpha)}))|}}. \tag{8.3}$$

where

$$R(\varphi, \varphi_1) \equiv \frac{1}{\omega_H}\int\limits_{\varphi-\varphi_1}^{\varphi} d\varphi' \mathbf{kv}(\varphi'), \langle \mathbf{kv}\rangle_\varphi = \frac{1}{2\pi}\int\limits_0^{2\pi} d\varphi' \mathbf{kv}(\varphi'), \tilde{\omega} = \omega + i0.$$

The summation is carried out over all the stationary points $\boldsymbol{\varphi}^{(\alpha)} = (\varphi^{(\alpha)}, \varphi_1^{(\alpha)})$; $\kappa(\boldsymbol{\varphi}^{(\alpha)}) = 1$ for the stationary points within the integration area $0 < \varphi^{(\alpha)} < 2\pi$, $0 < \varphi_1^{(\alpha)} < 2\pi$ and $\kappa(\boldsymbol{\varphi}^{(\alpha)}) = 1/2$ for the points located on its boundary,

$$\mathrm{s} = \mathrm{sign} R''_{\varphi\varphi_1}(\varphi^{(\alpha)}, \varphi_1^{(\alpha)}) = \nu_+(R''_{\varphi\varphi_1}) - \nu_-(R''_{\varphi\varphi_1}),$$

where $\nu_+(R''_{\varphi\varphi_1})$ and $\nu_-(R''_{\varphi\varphi_1})$ are the number of positive and negative eigenvalues of the matrix

$$R''_{\varphi\varphi_1} \equiv \frac{\partial^2 R(\varphi^{(\alpha)}, \varphi_1^{(\alpha)})}{\partial\varphi\partial\varphi_1}$$

respectively. Taking into account the weak dependence of the cross sectional area formed by the FS and the plane $p_H = const$ and the cyclotron mass on the projection of the momentum on the direction of the magnetic field p_H, the values m^* and ω_H^{-1} in Eq. (8.3) can be taken in the zero approximation in the two-dimensionality parameter η and $m^*\omega_H^{-1}$ can be taken out from under the integral sign. The dependence of the cyclotron frequency on the p_H should be considered only in the expression $k_x v_x/\omega_H$ in the exponent, provided that $\eta k v_F \simeq \omega_H$.

Assume that the magnetic field $\mathbf{H} = (H \sin \vartheta, 0, H \cos \vartheta)$ is located in the xz plane. Repeating the arguments of Sec. 7, we find that the maximal component of the tensor σ_{ij} is σ_{yy} which is proportional to $(kr_0)^{-1}$, Furthermore, the expansion of the component σ_{xj}, $j = x, y, z$ in powers of $(kr_0)^{-1}$ begins with the terms of higher order of smallness and the components $\sigma_{z\alpha}$, $\alpha = x, y, z$ are proportional to the powers of η. In the leading approximation in the small parameters $(kr_0)^{-1}$ and η we obtain from Eq. (8.2) the following dispersion relation:

$$\frac{k^2 c^2}{\omega^2} = \frac{4\pi i}{\omega} \sigma_{yy}. \tag{8.4}$$

In this approximation, the collective mode is transverse with the electric field polarized along the Y-axis. When taking into account the term proportional to $\sigma_{xy}\sigma_{yx}$, in the dispersion relation, the electric field of the wave defined by this equation has a small longitudinal component.

For the propagation of weakly damped waves it is necessary to satisfy the condition

$$|\omega - n\omega_H| > \max |\langle \mathbf{kv} \rangle_\varphi|. \tag{8.5}$$

Outside of the range of ω, $\mathbf{k}$, corresponding to this inequality, the integrand in Eq. (8.3) has a pole, and after integration with respect to p_H, the dispersion relation gets an imaginary part, responsible for the strong absorption of the wave. In layered conductors the drift velocity of electrons $\mathbf{v}_D = \langle \mathbf{v} \rangle_\varphi$ oscillates with changing the angle ϑ between the magnetic field and the normal to the layers. For certain directions of $\mathbf{H}$ with respect to the conductor layers, $\mathbf{v}_D$ is close to zero and there is no collisionless absorption, the wave attenuation is determined by electron scattering and the attenuation factor is proportional to frequency of collisions of the charge carriers. In this case, the existence of collective modes is possible even under strong spatial dispersion. In the range of ω and $\mathbf{k}$ such that $\mathbf{kv_m} \gg \omega_H$ and $\eta k v_F \simeq \omega_H$ where $\mathbf{v_m}$ is the maximum velocity in the direction of the wave vector $\mathbf{k}$, there are solutions of the dispersion relation (8.4) in the resonance region

$$\omega = n\omega_H + \Delta\omega, \tag{8.6}$$

where $n = 1, 2, 3 \ldots$, and $|\Delta\omega|$ is in the range $0 < |\Delta\omega| < \omega_H$.

To find the spectrum of the cyclotron modes explicitly, let us use Eq. (7.3) for the dispersion relation of the charge carriers. For the directions of $\mathbf{H}$ such that α equals one of the zeros $\alpha_i = (mv_F/p_0) \tan \vartheta_i$ of the Bessel function $J_0(\alpha)$, the average $\langle \mathbf{kv} \rangle_\varphi \simeq \eta^2$ and the dispersion relation takes the form

$$1+2\frac{(kr_0)^{-3}}{\pi\sin\phi}\left(\frac{\omega_p v_F}{\omega_H c}\right)^2\frac{\cos(\pi\tilde{\omega}/\omega_H)-\langle\sin R(\vartheta_i)\rangle_\beta}{\sin(\pi\tilde{\omega}/\omega_H)}=0 \qquad (8.7)$$

where

$$R(\vartheta_i)=\int_{-\pi/2}^{\pi/2}\frac{\mathbf{kv}(\varphi)}{\omega_B(\beta_i)}d\varphi=2\frac{k_x v_\perp}{\omega_H(\beta_i)}-\pi\eta\frac{k_z v_F}{\omega_H}\mathrm{S}_0(\alpha_i)\cos\beta_i$$

$$+\eta\frac{k_x v_F}{2\omega_H}\tan\vartheta_i\cos\beta_i\sum_{n=1}^{\infty}\frac{\mathrm{J}_{2n+1}(\alpha_i)}{n(n+1)(2n+1)}$$

$\mathrm{S}_0(\alpha)=(2/\pi)\int_0^{\pi/2}dt\sin(\alpha\cos t)$ – is the Struve function, $\beta_i=p_H/p_0\cos\vartheta_i$, $\omega_p=\sqrt{4\pi n_0 e^2/m}$ is the plasma frequency, and $n_0=p_0 m^2 v_F^2/2\pi\hbar^3$ is the density of charge carriers.

In the case of

$$\frac{1}{(kr_0)^3}\left(\frac{\omega_p v_F}{\omega_H c}\right)^2\gg 1,$$

which is easily realized in conductors with a charge carrier density of the order of one per atom, the solution of (8.4) can be written as

$$\omega=\omega_H\left(n-\frac{1}{2}+\frac{(-1)^n}{\pi}\arcsin\frac{1}{2\pi}\int_{-\pi}^{\pi}d\beta\sin\left(\int_{-\pi/2}^{\pi/2}\frac{\mathbf{kv}(\varphi)}{\omega_H(\beta)}d\varphi\right)\right) \qquad (8.8)$$

In the opposite limiting case,

$$\frac{1}{(kr_0)^3}\left(\frac{\omega_p v_F}{\omega_H c}\right)^2\ll 1$$

the spectrum of cyclotron waves is given by

$$\omega=n\omega_H\left(1-2\frac{(kr_0)^{-3}}{\pi\sin\phi}\left(\frac{\omega_p v_F}{\omega_H c}\right)^2\right.$$

$$\left.\times\left(1-(-1)^n\frac{1}{2\pi}\int_{-\pi}^{\pi}d\beta\sin\left(\int_{-\pi/2}^{\pi/2}\frac{\mathbf{kv}(\varphi)}{\omega_H(\beta)}d\varphi\right)\right)\right) \qquad (8.9)$$

In layered conductors for certain directions of an external magnetic field relative to the layers, the propagation of electromagnetic waves with frequencies in the vicinity of cyclotron resonance is possible for an arbitrary orientation of the vectors **k** and **H**. Due to the oscillatory dependence of the drift velocity on the angle between the magnetic field and the normal to the layers, transparency windows appear in a Q2D conductor for short-wave collective modes. Similar types of excitations in quasi-isotropic conductors are possible only when the wave vector is perpendicular to the external magnetic field.

Let us consider the special case when the wave vector $\mathbf{k} = (k, 0, 0)$ is orthogonal to the magnetic field $\mathbf{H} = (0, 0, H)$, directed along the normal to the layers. For an isotropic energy spectrum of electrons in the plane of the layers, we have $\varepsilon_{xx} = \varepsilon_{yy}$, $\varepsilon_{xz} = \varepsilon_{yz} = 0$ and the dispersion relation (8.1)

$$\det\left[k^2\delta_{ij} - k_ik_j - \frac{\omega^2}{c^2}\varepsilon_{ij}(\omega, \mathbf{k})\right]$$
$$= \left(k^2 - \frac{\omega^2}{c^2}\varepsilon_{zz}\right)\left(k^2\varepsilon_{xx} - \frac{\omega^2}{c^2}(\varepsilon_{xx}\varepsilon_{yy} - \varepsilon_{xy}\varepsilon_{yx})\right) = 0,$$

splits into two equations

$$\frac{k^2c^2}{\omega^2}\varepsilon_{xx} = \varepsilon_{xx}\varepsilon_{yy} + \varepsilon_{xy}^2, \quad \frac{k^2c^2}{\omega^2} = \varepsilon_{zz}, \tag{8.10}$$

the first of which was investigated above under the conditions of strong spatial dispersion for arbitrary orientations of the magnetic field and the wave vector. The second equation describes the transverse mode with the electric field polarized in the direction of the lowest conductivity. Let us find the spectrum of this wave [124]. In this case, the expressions for the components of the electron velocity are given by Eqs. (7.4) –(7.6) in which we set $\vartheta = 0$.

The interlayer conductivity σ_{zz} is given by

$$\sigma_{zz} = \frac{\eta^2\omega_p^2}{8\pi^2\omega_H}\int_{-\pi}^{\pi} d\beta \sin^2\beta\left(1 - \exp\left(2\pi i\frac{\tilde{\omega}}{\omega_H}\right)\right)^{-1}$$
$$\times \int_0^{2\pi}\int_0^{2\pi} d\varphi d\varphi_1 \exp\left(iR(\varphi, \varphi_1) - i\frac{\tilde{\omega}}{\omega_H}\varphi_1\right), \tag{8.11}$$

where $R(\varphi, \varphi_1) = -(kv_\perp/\omega_H)\sin\varphi + (kv_\perp/\omega_H)\sin(\varphi - \varphi_1)$, $v_\perp = v_F\sqrt{1+\varepsilon\cos\beta} = v_F(1+\frac{1}{2}\varepsilon\cos\beta)$ is the velocity in plane of the layers, $\varepsilon = \eta v_F p_0/\varepsilon_F$, $\beta = p_z/p_0$, $p_z = p_H$.

After integration with respect to φ and φ_1 we obtain

$$\sigma_{zz} = \frac{i\eta^2\omega_p^2}{4\pi\omega_H}\int_{-\pi}^{\pi} d\beta \frac{\mathrm{J}_{\tilde{\omega}/\omega_H}(2kv_\perp/\omega_H)\mathrm{J}_{-\tilde{\omega}/\omega_H}(2kv_\perp/\omega_H)}{\sin(\pi\tilde{\omega}/\omega_H)} \tag{8.12}$$

In the case of strong spatial dispersion $kr_0 \gg 1$, it is possible to use the asymptotic representation of the Bessel function for large values of the argument and to integrate with respect to β. As a result, the dispersion equation takes the form

$$\frac{k^2c^2}{\omega^2} = 1 - \eta^2\frac{\omega_p^2}{\omega_H\omega}\frac{1}{kr_0}\left(\cot\frac{\pi\tilde{\omega}}{\omega_H} + \frac{2J_1(\varepsilon kr_0)}{\varepsilon kr_0}\frac{\sin 2kr_0}{\sin(\pi\tilde{\omega}/\omega_H)}\right), \tag{8.13}$$

Under the condition

$$\frac{\eta^2}{(kr_0)^3}\left(\frac{\omega_p v_F}{\omega_H c}\right)^2 \ll 1,$$

which is almost always fulfilled in strongly anisotropic organic conductors for sufficiently large values of kr_0, we obtain the following expression for the frequency of the cyclotron wave from Eq. (8.13)

$$\omega(k) = n\omega_H\left\{1 - \frac{\eta^2}{\pi(kr_0)^3}\left(\frac{\omega_p v_F}{\omega_H c}\right)^2\left(1 + (-1)^n\frac{2J_1(\varepsilon kr_0)}{\varepsilon kr_0}\sin 2kr_0\right)\right\},$$
$$n = 1, 2, 3, \ldots. \tag{8.14}$$

In this equation, the factor with the Bessel function $2J_1(\varepsilon kr_0)/\varepsilon kr_0$ appears after integration with respect to β, due to the dependence of the electron velocity in the plane of the layers $v_\perp$ on the momentum projection on the magnetic field direction. In the case of extremely low values of the anisotropy parameter $v_\perp = v_F$ and $\varepsilon kr_0 \ll 1$, we obtain $2J_1(\varepsilon kr_0)/\varepsilon kr_0 \approx 1$.

As follows from Eqs. (8.8), (8.9) and (8.14) under the conditions of strong spatial dispersion, the cyclotron wave frequencies are oscillatory functions of the wave vector.

9. Collective modes in Q1D organic conductors

The most common examples of conductors with strongly anisotropic FS of QlD type in the form of a pair of slightly warped sheets are the so-called Bechgaard salts $(TMTSF)_2X$ (X stands for a set of various anions). Usually this electron energy spectrum can be presented in the form Eq. (2.7)

$$\varepsilon(\mathbf{p}) = v_F(|p_x| - p_F) + B\cos\frac{p_y}{p_2} + C\cos\frac{p_z}{p_3}.$$

In the absence of quantization of electron energy levels in a magnetic field, for the frequencies of alternating electromagnetic field ω less than $C/\hbar$, we can describe the kinetic properties of the conductor with the aid of the quasiclassical approximation. In the case when the magnetic field $\mathbf{H} = (0, H\sin\vartheta, H\cos\vartheta)$ is perpendicular to the chain direction, the components of electron velocity are given by

$$v_x = \mathrm{sign}(p_x)v_F, \quad v_y = \mathrm{sign}(p_x)v_2\sin\Omega t,$$

$$v_z = v_3\sin\left(\frac{p_H}{p_3\cos\vartheta} - \mathrm{sign}(p_x)\alpha\Omega t\right), \quad \Omega = (|e|v_F H_0/cp_2)\cos\vartheta, \quad (9.1)$$

here $\alpha = (p_2/p_3)\tan\vartheta$, $v_2 = B/p_2$ and $v_3 = C/p_3$ are characteristic electron speeds in the plane perpendicular to the conducting chain, The values $\mathrm{sign}(p_x) = \pm 1$ correspond to different sheets of FS.

The electrical conductivity tensor can be written as

$$\sigma_{ij}(\omega, \mathbf{k}) = \frac{2|e|^3 H}{(2\pi\hbar)^3 c}\sum_{\mathrm{sign}(p_x)=\pm 1}\int dp_H \int_0^{2\pi/\Omega} dt v_i(t)$$

$$\times \int_{-\infty}^{t} dt v_j(t')\exp\left(i\tilde{\omega}(t-t') - i\int_{t'}^{t} dt''\mathbf{k}\mathbf{v}(t'')\right), \quad (9.2)$$

here $\tilde{\omega} = \omega + i\tau^{-1}$. As the variables in the momentum space we have chosen the integrals of motion ε, p_H and the time t of motion of an electron in a magnetic field along the trajectory defined by the quasi-classical dynamics equations.

Consider the case when wave vector $\mathbf{k} = (0, k\sin\phi, k\cos\phi)$ is orthogonal to the direction of most conductivity. For the energy spectrum and the geometry of the problem under consideration, the tensor components σ_{ij} take the form [125, 126]

$$\sigma_{xx} = \frac{\omega_p^2}{2\pi\Omega}\int_0^\infty d\varphi e^{i\frac{\tilde{\omega}}{\Omega}\varphi} J_0\left(2Y\sin\frac{\alpha\varphi}{2}\right) J_0\left(2X\sin\frac{\varphi}{2}\right), \tag{9.3}$$

$$\sigma_{yy} = \frac{\omega_p^2}{4\pi\Omega}\left(\frac{v_2}{v_F}\right)^2\int_0^\infty d\varphi e^{i\frac{\tilde{\omega}}{\Omega}\varphi} J_0\left(2Y\sin\frac{\alpha\varphi}{2}\right)$$

$$\times\left(J_0\left(2X\sin\frac{\varphi}{2}\right)\cos\varphi - J_2\left(2X\sin\frac{\varphi}{2}\right)\right), \tag{9.4}$$

$$\sigma_{zz} = \frac{\omega_p^2}{4\pi\Omega}\left(\frac{v_3}{v_F}\right)^2\int_0^\infty d\varphi e^{i\frac{\tilde{\omega}}{\Omega}\varphi} J_0\left(2X\sin\frac{\varphi}{2}\right)$$

$$\times\left(J_0\left(2Y\sin\frac{\alpha\varphi}{2}\right)\cos\alpha\varphi - J_2\left(2Y\sin\frac{\alpha\varphi}{2}\right)\right), \tag{9.5}$$

$$\sigma_{yz} = \sigma_{zy} = -\frac{\omega_p^2}{2\pi\Omega}\frac{v_2 v_3}{v_F}\int_0^\infty d\varphi e^{i\frac{\tilde{\omega}}{\Omega}\varphi} J_1\left(2Y\sin\frac{\alpha\varphi}{2}\right)$$

$$\times J_1\left(2X\sin\frac{\varphi}{2}\right)\cos\frac{\varphi}{2}\cos\frac{\alpha\varphi}{2} \tag{9.6}$$

$$\sigma_{xy} = \sigma_{yx} = \sigma_{xz} = \sigma_{zx} = 0$$

where $X = k_y v_2/\Omega$, $Y = k_z v_3/(\alpha\Omega)$, $J_n(x)$ is the n-th order Bessel function. For values of the transfer integral $A \sim 0.5\,\mathrm{eV}$, the frequency $\omega_p = (4e^2 p_2 p_3 v_F/\hbar^3)^{1/2}$ is of the order of $10^{15}\,s^{-1}$.

It is easy to see that oscillations of the components v_y, v_z of electron velocity lead to the resonances in high-frequency conductivity. Expanding the Bessel functions in Eqs. (9.3)–(9.6) into Fourier series with respect to φ and $\alpha\varphi$

$$J_0(2Z\sin(\psi/2)) = \sum_{n=-\infty}^{\infty} J_n^2(Z)\exp(in\psi),$$

$$J_2(2Z\sin(\psi/2)) = \sum_{n=-\infty}^{\infty} J_{1-n}(Z)J_{1+n}(Z)\exp(in\psi),$$

$$Z = \{X, Y\}, \quad \psi = \{\varphi, \alpha\varphi\},$$

and integrating over φ, we obtain the following expressions for the diagonal components of conductivity tensor [126, 127]

$$\sigma_{xx} = \frac{i\omega_p^2}{2\pi} \sum_{n,m=-\infty}^{\infty} \frac{\mathrm{J}_n^2(\mathrm{X})\mathrm{J}_m^2(\mathrm{Y})}{\tilde{\omega} - n\Omega - \alpha m\Omega}, \tag{9.7}$$

$$\sigma_{yy} = \frac{i\omega_p^2}{4\pi}\left(\frac{v_2}{v_F}\right)^2 \sum_{n,m=-\infty}^{\infty} \frac{\mathrm{J}_m^2(\mathrm{Y})(\mathrm{J}_{n-1}^2(\mathrm{X}) + \mathrm{J}_{n+1}^2(\mathrm{X}) + 2\mathrm{J}_{n-1}(\mathrm{X})\mathrm{J}_{n+1}(\mathrm{X}))}{\tilde{\omega} - n\Omega - \alpha m\Omega}, \tag{9.8}$$

$$\sigma_{zz} = \frac{i\omega_p^2}{4\pi}\left(\frac{v_3}{v_F}\right)^2 \sum_{n,m=-\infty}^{\infty} \frac{\mathrm{J}_n^2(\mathrm{X})(\mathrm{J}_{m-1}^2(\mathrm{Y}) + \mathrm{J}_{m+1}^2(\mathrm{Y}) + 2\mathrm{J}_{m-1}(\mathrm{Y})\mathrm{J}_{m+1}(\mathrm{X}))}{\tilde{\omega} - n\Omega - \alpha m\Omega}. \tag{9.9}$$

In the case when mean free time τ is sufficiently large i.e. $\Omega\tau \gg 1$, the local maximums of the high-frequency conductivity and microwave absorbing occur at the condition

$$\omega - n\Omega - \alpha m\Omega = 0. \tag{9.10}$$

However, the resonances at the frequencies $\omega = \alpha m\Omega$ resulting from electron motion in z-direction may show up only for sufficiently short-wavelength electromagnetic field, when $\mathrm{Y}^2\alpha\Omega\tau$ is comparable with unity.

For $Y \ll 1$, Eqs. (9.3)–(9.6) can be simplified. For example, we can let $\mathrm{J}_0(2\mathrm{Y}\sin(\alpha\varphi/2)) = 1$ in Eq. (9.3) integrating the result over φ, we find

$$\sigma_{xx} = \frac{i\omega_p^2}{2\Omega} \frac{\mathrm{J}_{\tilde{\omega}/\Omega}(\mathrm{X})\mathrm{J}_{-\tilde{\omega}/\Omega}(\mathrm{X})}{\sin(\pi\tilde{\omega}/\Omega)}. \tag{9.11}$$

Formula (9.3) also can be simplified at arbitrary values of parameter Y, if the wave vector $\mathbf{k} = (0, 0, k)$ is parallel to the direction of lowest conductivity

$$\sigma_{xx} = \frac{i\omega_p^2}{2\alpha\Omega} \frac{\mathrm{J}_{\tilde{\omega}/(\alpha\Omega)}(\mathrm{Y})\mathrm{J}_{-\tilde{\omega}/(\alpha\Omega)}(\mathrm{Y})}{\sin(\pi\tilde{\omega}/(\alpha\Omega))}. \tag{9.12}$$

In the collisionless limit $\tau \to \infty$ the high-frequency conductivity may become non-dissipative; that results in the possibility of appearance of weakly damped collective modes in the conductors with strongly anisotropic FS of QlD type.

For the chosen approximation for electron energy spectrum in the case when the electric current flows along the conducting chain, Eq. (8.1) can be reduced to the form

$$\frac{k^2c^2}{\omega^2} = \frac{4\pi i}{\omega}\sigma_{xx}, \tag{9.13}$$

This equation determines transverse electromagnetic waves with the electric field polarized along the x-axis and frequencies

$$\omega = n\Omega + \Delta\omega, \quad n = 1, 2, 3\ldots, \tag{9.14}$$

where $\Delta\omega$ is within the range $0 < |\Delta\omega| < \Omega$. The spectrum can be presented in analytical form in short- and long-wavelength limits. For sufficiently large $\mathrm{X} \gg 1$ and at the same time $\mathrm{Y} \ll 1$, we can simplify Eq. (9.13), using in Eq. (9.3) the asymptotic representation of the Bessel functions

$$1 + \frac{\omega\beta}{\omega_0(kr_0)^3 \sin\phi} \frac{\cos(\pi\tilde{\omega}/\Omega) + \sin 2\mathrm{X}}{\sin(\pi\tilde{\omega}/\Omega)} = 0, \tag{9.15}$$

here $\beta = (\omega_p v_2)^2/(\omega_0 c)^2$, $\omega_0 = (|e|v_F H)/(cp_2)$, $r_0 = v_2/\omega_0$. At small values of the parameter $\xi = \beta/(k^3 r_0^3 \sin\varphi) \ll 1$ the eigen frequencies are close to the resonance frequency and its harmonics

$$\omega = n\Omega\left(1 - \frac{\xi}{\pi}(1 + (-1)^n \sin 2\mathrm{X})\right) - i\tau^{-1}\left(1 + \frac{\xi}{\pi}(1 + (-1)^n \sin 2\mathrm{X})\right). \tag{9.16}$$

The deviation of ω from $n\Omega$ oscillates as $\sin 2\mathrm{X}$ and decreases with k as k^{-3}.

In the opposite limiting case $\xi \gg 1$, the approximate solutions of the Eq. (9.15)

$$\omega = \omega^{(0)} + \frac{\Omega^2}{\pi}(\xi\omega^{(0)} - (-1)^n\Omega \tan 2\mathrm{X})^{-1},$$

$$\omega^{(0)} = \Omega\left(n - \frac{1}{2} - \frac{(-1)^n}{\pi}\arcsin(\sin 2\mathrm{X})\right) - i\tau^{-1}, \tag{9.17}$$

are essentially distinct from $n\Omega$.

Under weak spatial dispersion $\mathrm{X} \ll 1$, formula (9.11) can be expanded in rapidly decreasing power series. In the first order in X^2, the conductivity takes form

$$\sigma_{xx} = \frac{i\omega_p^2}{2\pi\tilde{\omega}}\left(1 - \frac{k_y^2 v_2^2}{2(\Omega^2 - \tilde{\omega}^2)}\right) \tag{9.18}$$

The dispersion equation can be easily solved. The frequency of long-wavelength mode is

$$\omega = \Omega - \Omega\frac{X^2}{4}\left(1 - \frac{1}{2}\left(\frac{kc}{\omega_p}\right)^2\right) - i\tau^{-1}\left(1 + \frac{X^2}{8}\left(\frac{kc}{\omega_p}\right)^2\right). \tag{9.19}$$

For arbitrary values of X and Y the solution of transcedental equation (9.13) cannot be obtained in analytical form. The numerical calculations of the spectra of the first five collective modes in the collisionless limit $\tau^{-1} \to 0$ at various orientations of magnetic field and wave vector are presented in Fig. 14. The weakly damped waves vanish at $\omega = n\Omega$ due to strong cyclotron absorption.

The case when the wave vector $\mathbf{k} = (0, 0, k)$ is parallel to z-axis, is most favorable for appearance of the resonances resulting from the periodic movement of conduction electrons in the direction of least conductivity direction. At $Y \ll 1$ we can solve Eq. (9.13) using the asymptotic expansion of formula (9.12). The frequency of the first eigenmode can be written as

$$\omega = \alpha\Omega - \alpha\Omega\frac{Y^2}{4}\left(1 - \frac{1}{2}\left(\frac{kc}{\omega_p}\right)^2\right) - i\tau^{-1}\left(1 + \frac{Y^2}{8}\left(\frac{kc}{\omega_p}\right)^2\right), \tag{9.20}$$

The dispersion of the waves caused by electron oscillation in z-direction at various orientations of magnetic field is shown in Fig. 15.

If $\vartheta = 0$ and $\phi = 0$ i.e. $\mathbf{H} = (0,\ 0,\ H) \parallel \mathbf{k}$, the conductivity

$$\sigma_{xx} = \frac{i\omega_p^2}{4\pi\sqrt{\tilde{\omega}^2 - k^2 v_3^2}}$$

is independent on **H** and the spectrum of weakly damped collective modes lies above ω_p.

The necessary criterion for propagation of weakly damped resonance modes, as well as for other high-frequency resonance phenomena in a magnetic field, is $\omega_0\tau \gg 1$. In the numerical calculations presented above we have assumed $\tau^{-1} \to 0$. We illustrate with the aid of Fig. 16, the influence of electron collisions on the wave process.

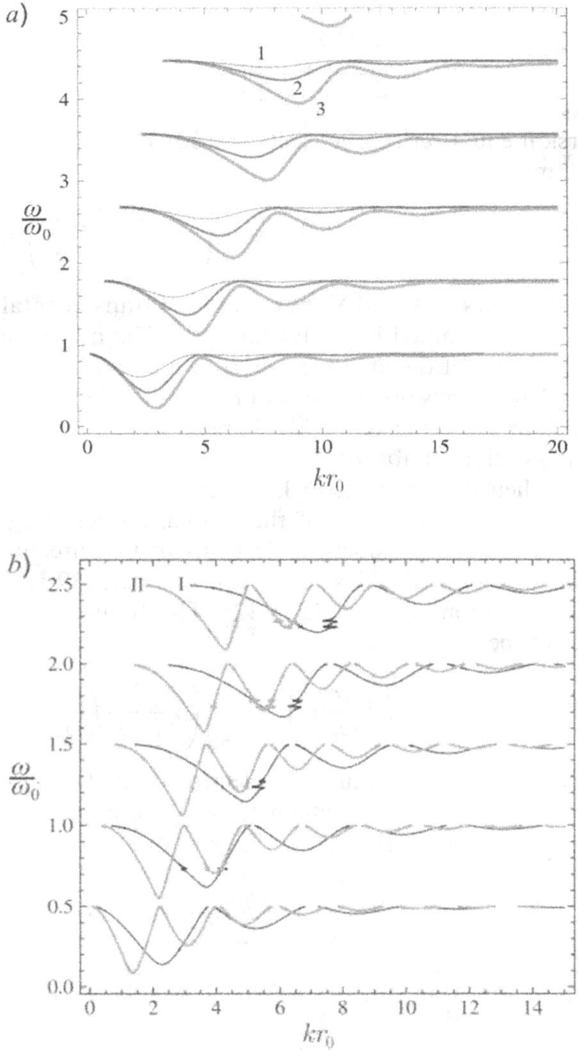

Fig. 14. Spectra of the first five collective modes at various values of ϑ and ϕ; a) $\vartheta = \arctan(1/2)$, $\phi = \pi/4$, curves 1, 2, 3, correspond to $\beta = 5$, 20, 100 respectively, b) $\vartheta = \pi/3$, $\beta = 50$, curves I, II correspond to $\phi = \pi/6$ and $\phi = \pi/3$. Here $\tau^{-1} \to 0$, $p_2/p_3 = 2$, $C/B = 1/25$.

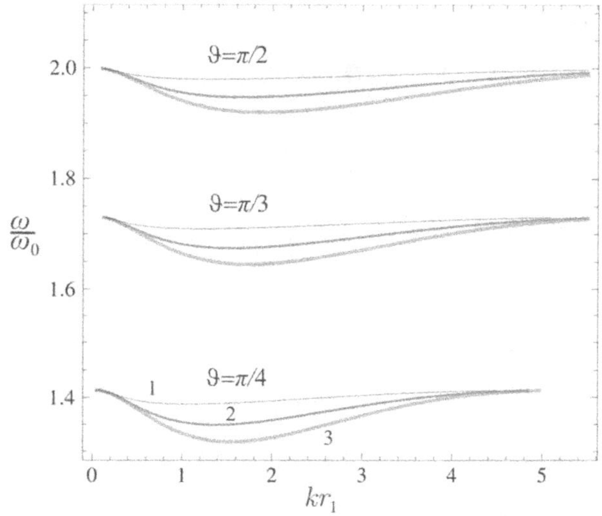

Fig. 15. Spectra of the first collective modes at $\phi = 0$, $\vartheta = \pi/4$, $\pi/3$, $\pi/2$, $\tau^{-1} \to 0$, $p_2/p_3 = 2$; curves 1, 2, 3 correspond to $\beta_1 = 0.1$, 0.3, 0.5, respectively. Here $r_1 = v_3/\omega_0$, $\beta_1 = (\omega_p v_3)^2/(\omega_0 v_3)^2$.

The weakly damped collective modes do not exist in the vicinity of the order of τ^{-1} near the resonance frequencies $n\Omega$ because of strong absorption. For the values of ω so that $|\omega - n\Omega| < \tau^{-1}$, the conductivity $\sigma_{xx}(\omega, \mathbf{k})$ gains a large real part, which is responsible for the strong attenuation of the wave.

The collective modes with frequencies near resonance frequency and its harmonics can occur only under the condition of non-local link with current density and electrical field. The dispersion effects are more essential for the existence of high-order harmonics. As it follows from formulas (9.16), (9.17) and Fig. 13, the dependence of eigen frequencies on a wave vector has a pronounced oscillatory character.

Let us consider the special case when wave vector $\mathbf{k} = (0, k, 0)$ is directed along y-axes. The dispersion equation (8.1) is factored

$$D = \left(k^2 - \frac{\omega^2}{c^2}\varepsilon_{xx}(\omega, \mathbf{k})\right)\left(-\frac{\omega^2}{c^2}\varepsilon_{yy}(\omega, \mathbf{k})\right)\left(k^2 - \frac{\omega^2}{c^2}\varepsilon_{zz}(\omega, \mathbf{k})\right) = 0 \tag{9.21}$$

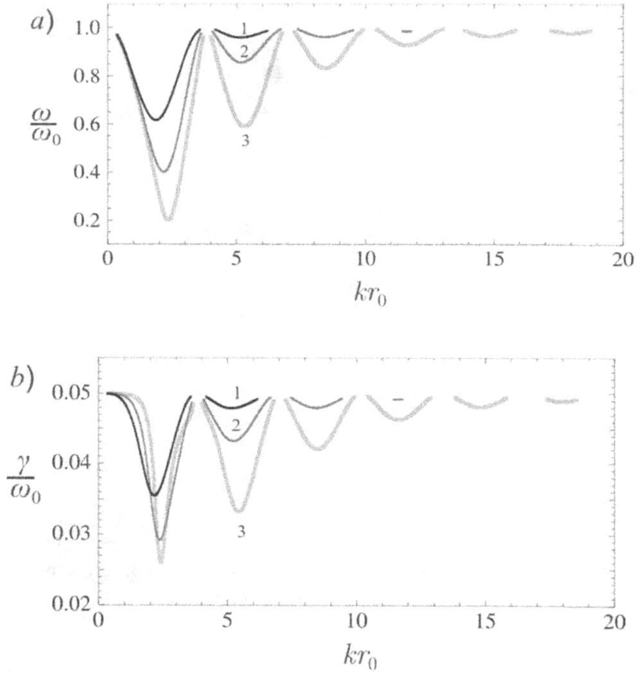

Fig. 16. The frequency ω (a) and damping decrement γ (b) of the first collective mode at $(\omega_0\tau)^{-1} = 0.05$, $\vartheta = 0$, $\phi = \pi/2$, curves 1, 2, 3 correspond to $\beta = 5$, 20, 100 respectively.

and splits into three equations. The first equation describes the transverse mode with the electric field polarized in the direction of maximum conductivity, and was studied in detail above for the arbitrary orientations of the magnetic field and the wave vector. The second equation has weakly damped solutions in the frequency range $\omega > (B/\varepsilon_F)^2\omega_p$ and determines the longitudinal plasma oscillations in the direction y. The third equation

$$k^2 - \frac{\omega^2}{c^2}\varepsilon_{zz}(\omega, \mathbf{k}) = 0 \tag{9.22}$$

describes the collective mode with the electric field polarized in the direction of lowest conductivity(z-axes). In the case $k_z = 0$, the integral expression for conductivity component (9.5) can be reduced to the form

$$\sigma_{zz} = \frac{i\omega_p^2}{4\Omega}\left(\frac{v_3}{v_F}\right)^2$$

$$\times \left[\frac{J_{(\tilde{\omega}-\alpha\Omega)/\Omega}(X)J_{-(\tilde{\omega}-\alpha\Omega)/\Omega}(X)}{\sin\left(\pi\dfrac{\tilde{\omega}-\alpha\Omega}{\Omega}\right)} + \frac{J_{(\tilde{\omega}+\alpha\Omega)/\Omega}(X)J_{-(\tilde{\omega}+\alpha\Omega)/\Omega}(X)}{\sin\left(\pi\dfrac{\tilde{\omega}+\alpha\Omega}{\Omega}\right)} \right]. \tag{9.23}$$

Under weak spatial dispersion $X \ll 1$, formula (9.23) can be expanded into power series in X^2 and equation becomes algebraic. In the main approximation in parameter X^2, low-frequency mode dispersion is given by

$$\omega = \frac{\alpha\Omega}{\sqrt{2}}\frac{kcv_F}{\omega_p v_3}. \tag{9.24}$$

In short-wavelength limit $X \gg 1$, we can simplify Eq. (9.22), using asymptotic representation of Bessel functions in the formula (9.23) in the form of trigonometrical functions. At small values of parameter $(1/\pi X^3)(\omega_p v_2 v_3)^2/(v_F \Omega c)^2 \ll 1$ eigen frequencies are close to resonance frequencies $\omega = (n \pm \alpha)\Omega$

$$\omega = (n \pm \alpha)\Omega\left(1 - \frac{1}{\pi X^3}\left(\frac{\omega_p v_2}{\Omega c}\right)^2\left(\frac{v_3}{v_F}\right)^2(1-(-1)^n \sin 2X)\right) \tag{9.25}$$

The numerical calculations of spectra of collective modes for arbitrary values of wave vector in the limiting case of large relaxation times are presented on Fig. 17.

As can be seen from the figure, weakly damped waves disappear when the wave frequency is close to the resonant frequencies $\omega = (n \pm \alpha)\Omega$ because of to strong cyclotron absorption.

In the case, when wave vector $\mathbf{k} = (0, 0, k)$ is parallel to the z-axis, the dispersion equation will be transformed to the form

$$D = \left(k^2 - \frac{\omega^2}{c^2}\varepsilon_{xx}(\omega, \mathbf{k})\right)\left(k^2 - \frac{\omega^2}{c^2}\varepsilon_{yy}(\omega, \mathbf{k})\right)\left(-\frac{\omega^2}{c^2}\varepsilon_{zz}(\omega, \mathbf{k})\right) = 0 \tag{9.26}$$

a)

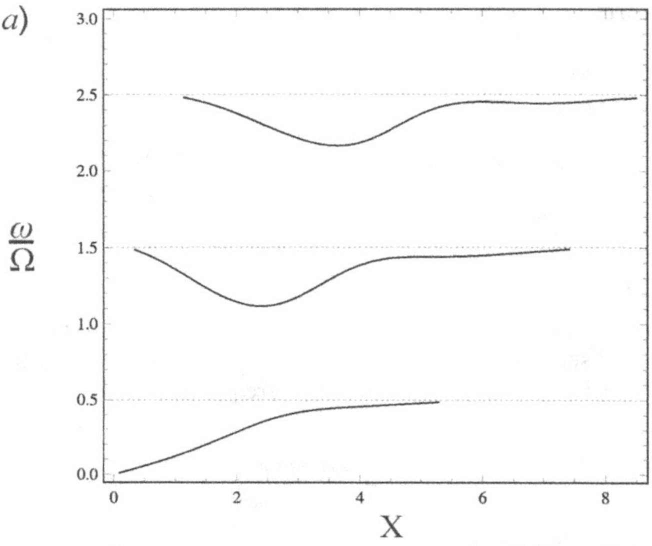

b)

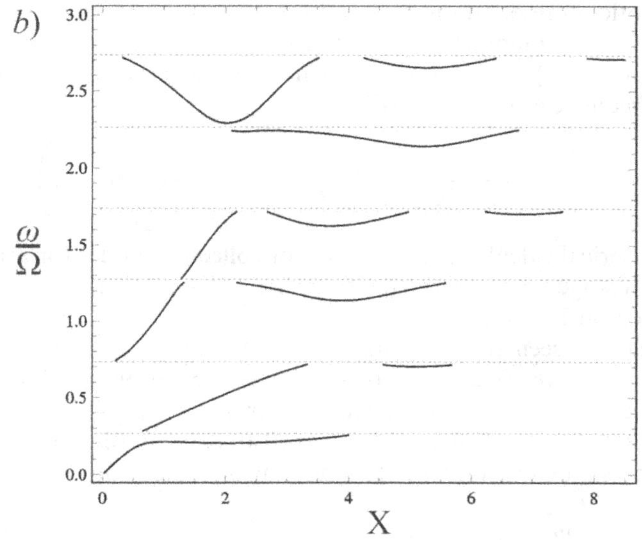

Fig. 17. Spectra of the collective modes a $\mathbf{k} = (0, k, 0)$, $(\omega_p v_2 v_3 / \alpha \Omega v_F c)^2 = 10$, $(\Omega\tau)^{-1} = 0.01$; a) $\alpha = 1/2$, b) $\alpha = \sqrt{3}$. The dotted lines correspond to the resonance frequencies $\omega = (n \pm \alpha)\Omega$.

Equation

$$k^2 - \frac{\omega^2}{c^2}\varepsilon_{yy}(\omega, \mathbf{k}) = 0 \tag{9.27}$$

describes the collective mode with the electric field polarized in y-direction. After calculation of the integral in the formula (9.4) for conductivity components σ_{yy} we obtain

$$\sigma_{yy} = \frac{i\omega_p^2}{4\alpha\Omega}\left(\frac{v_2}{v_F}\right)^2 \times \left[\frac{\mathrm{J}_{(\tilde{\omega}-\Omega)/\alpha\Omega}(\mathrm{Y})\mathrm{J}_{-(\tilde{\omega}-\Omega)/\alpha\Omega}(\mathrm{Y})}{\sin\left(\pi\frac{\tilde{\omega}-\Omega}{\alpha\Omega}\right)} + \frac{\mathrm{J}_{(\tilde{\omega}+\Omega)/\alpha\Omega}(\mathrm{Y})\mathrm{J}_{-(\tilde{\omega}+\Omega)/\alpha\Omega}(\mathrm{Y})}{\sin\left(\pi\frac{\tilde{\omega}+\Omega}{\alpha\Omega}\right)}\right] \tag{9.28}$$

At $\mathrm{Y} \ll 1$, it is possible to simplify the equation (9.27) by means of asymptotic expansion of the formula (9.28) in a series on degrees Y. As a result we find the dispersion relation for the low-frequency mode

$$\omega = \frac{\Omega}{\sqrt{2}}\frac{kcv_F}{\omega_p v_2}, \tag{9.29}$$

The spectra of the waves propagating in the direction of the smallest conductivity are represented in Fig. 18.

The weakly damped eigenmodes represent the collective excitations of a Bose type in electron plasma of the solid. The physical nature of the resonance electromagnetic modes in highly anisotropic conductors with electron energy spectrum of Q1D type is connected with the periodic movement of conduction electrons in strong magnetic field $\mathbf{H} = (0, H\sin\vartheta, H\cos\vartheta)$ almost without collisions $(\omega_0\tau \gg 1)$ across the warped FS sheets. The components v_y and v_z of electron velocity oscillate with frequencies Ω and $\Omega_1 = (p_2/p_3)\tan\vartheta\,\Omega$, respectively, generating resonances in the kinetic coefficients of the conductors. So, resonances in high-frequency conductivity can occur at two resonance frequencies and its harmonics. However, the resonances at the frequencies $\omega = n\Omega_1$ resulting from electron movement along the least conductivity direction may show up only for sufficiently short-wavelength electromagnetic field when $(kv_3/\Omega_1)^2\Omega_1\tau$ is comparable with unity. The spatial dispersion is a necessary condition for the

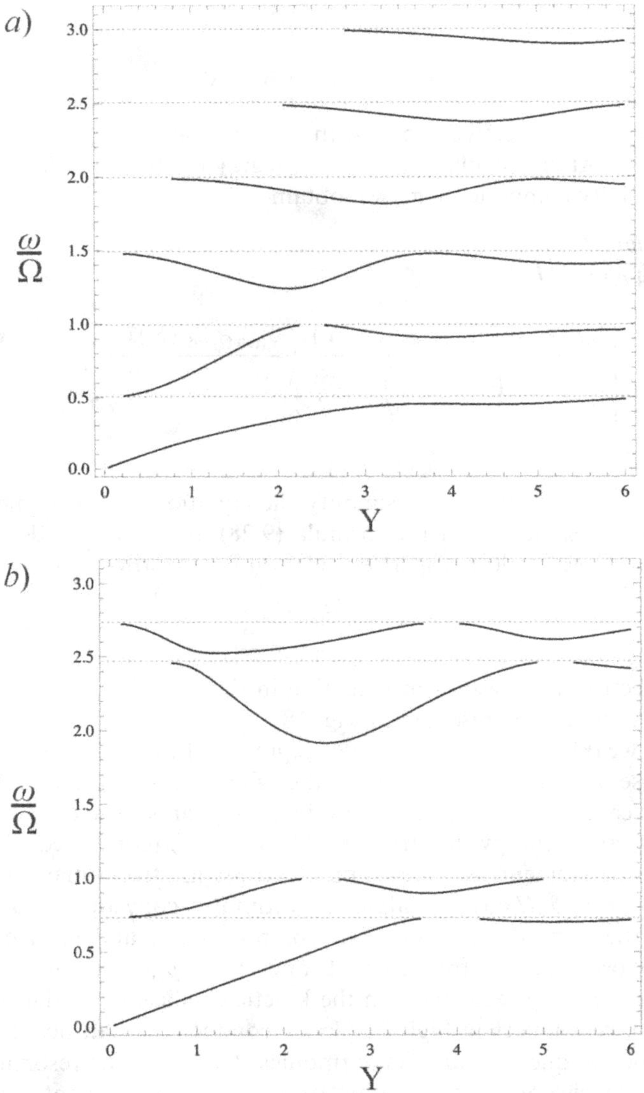

Fig. 18. Spectra of the first collective modes at $\mathbf{k} = (0, 0, k)$, $(\omega_p v_2 v_3/\alpha\Omega v_F c)^2 = 10$, $(\Omega\tau)^{-1} = 0.01$; a) $\alpha = 1/2$, b) $\alpha = \sqrt{3}$. The dotted lines correspond to the resonance frequencies $\omega = (\alpha n \pm 1)\Omega$.

existence of electromagnetic modes with frequencies ω near the resonances $n\Omega_r$, moreover, the dispersion effects become more essential with increase in ω. The influence of electron collisions on the wave process results in disappearance of weakly damped collective modes in the vicinity of the order of τ^{-1} near the resonance frequencies due to strong absorption. For this reasons only the lowest order modes resulting from the periodic movement of conduction electrons in the direction of least conductivity may occur mostly when the wave vector is parallel to z-axis. In the case when the orientation of the wave vector is not too close to the least conductivity direction, the eigen frequencies can be presented as $\omega = n\Omega + \Delta\omega$, $n = 1, 2, 3\ldots$, where $\Delta\omega$ has pronounced oscillatory dependence on k and decreases as k^{-3} at large values of k. The analytical expressions for $\Delta\omega$ can be obtained in a number of limiting cases.

10. The kinetic equations in Q2D electron Fermi-liquid

The properties of electron plasma in solids are specified by the presence of the Fermi-liquid interaction between electrons. The average energy of the Coulomb interaction of electrons is of the order of magnitude of their average kinetic energy. Therefore, to successfully explain the wide range of phenomena, strong electron-electron interaction consistently must be taken into account. In 1956, Landau developed the necessary method, known as the Fermi-liquid theory [128], and used this theory to describe the properties of a neutral Fermi liquid ^{3}He. Silin [129, 130] extended the Landau theory to the case of long-range Coulomb forces in corresponding electron liquid in metals. According to Landau – Silin theory, the excited state of the system of interacting fermions with the excitation energy $\varepsilon(\mathbf{p}) - \varepsilon_F$ is small compared with the Fermi energy ε_F and can be described in terms of simple one-particle excitations – quasi-particles that obey Fermi statistics. Each quasi-particle has a quasi-momentum, and their number is equal to the number of particles. Excited states arise when the electron goes from the occupied state with energy smaller than Fermi energy in the free state. Due to the Pauli principle, the considered one-particle excitations can exist long enough even in Fermi systems with strong interaction. If one electron is transferred to an excited state $\varepsilon(\mathbf{p})$ near the Fermi level $\varepsilon(\mathbf{p}) - \varepsilon_F \ll \varepsilon_F$, it is clear that this state is not the true stationary state of many-electron system. As a result of the Coulomb interaction with other electrons, this one-particle excitation should fade over time due to transitions to other states. However, regardless of the specific features of the Fermi liquid, the decay rate should be low for quasi-particles close to the Fermi surface, because the electrons occupy almost all of the states in which this excitation can relax. A quantum-mechanical calculation shows that the lifetime of such one-particle state is proportional to $(\varepsilon(\mathbf{p}) - \varepsilon_F)^{-2}$ Therefore, single-particle excitations near the ground state are almost stationary states of a system of interacting fermions.

In contrast to the ideal Fermi gas, the energy of the individual quasi-particle in a Fermi liquid depends on the states of the surrounding particles due to the significant self-consistent interaction. The total energy of the system is not equal to additive sum of the energies of individual quasi-particles, and represents some nonlinear functional of their distribution function $f(\mathbf{p}, \mathbf{r}, t)$. In this case, the energy of quasi-particle is defined as the variational derivative of the total energy of the system with respect to $f(\mathbf{p}, \mathbf{r}, t)$. The Fermi-liquid theory of the interaction between the quasi-particles can be described with the aid of the Landau correlation function, which is the second variational derivative of the energy density of the system with respect to distribution function. It is assumed that the Landau function is a definite quantity and it is used to calculate the characteristics of electron liquid. Information about the correlation function, i.e., the values of a number of phenomenological parameters used in this theory, are determined from experimental data. This method is similar to the problem of reconstructing the form of the Fermi surface from the results of measurements of the magnetic oscillations of kinetic and thermodynamic characteristics of the metal in a quantizing magnetic field.

A significant qualitative difference between the properties of the electron Fermi liquid and the electron gas is shown in rather high-frequency processes. Such processes include, in particular, high frequency resonances and propagation of collective modes in the presence of an external magnetic field. An example of collective excitations which are absent in gas of non-interacting particles and caused by the exchange interaction between the electrons are paramagnetic spin waves predicted by Silin [130] and experimentally detected by Dunifer and Shultz [131] in alkali metals. Studies of cyclotron resonance and other high frequency phenomena can provide information on the band structure of solids and constants of electron-electron interaction, which cannot be obtained by other means.

The specific of the quasi-two-dimensional electron energy spectrum and of the Fermi-liquid interaction in layered conductors leads to the occurrence of weakly-damping eigen oscillations of the electron and spin densities which are absent in a quasi-isotropic metal. As the interaction between charge carriers inside conduction layers considerably exceeds the interaction between quasi-particles belonging to different layers, the Landau correlation function can be expanded in asymptotic series in powers of the small parameter describing the anisotropy of the electron energy spectrum; the leading term of the expansion being independent on the momentum projection on the normal to the layers [132, 133]. As a result the integral equations for the renormalized distribution function for charge carriers and spin density can be obtained as a general form of the Landau correlation function.

The kinetic properties of the system of fermions in an electromagnetic field are described by means of the kinetic equation for the density matrix and the Maxwell system of equations. For perturbations that vary slightly at a distance of the order of the de Broglie wavelength, quasi-particles have certain values both coordinate and momentum and the density matrix $\hat{\rho}(\mathbf{p}, \mathbf{r}, \boldsymbol{\sigma}, t)$ represents an operator in the space of spin variables and function of the coordinates and of the momenta. In this case, the quasi-classical distribution function is defined here by the relation $f(\mathbf{p}, \mathbf{r}, t) = \mathrm{Tr}_\sigma\, \hat{\rho}(\mathbf{p}, \mathbf{r}, \boldsymbol{\sigma}, t)$ here $\boldsymbol{\sigma}$ are Paulie's matrices, Tr_σ is trace over the spin states. . In the presence of strong external magnetic field for the validity of the quasi-classical approximation it is necessary that the quantization of the energy levels of charge carriers has no significant effect on the kinetic and thermodynamic characteristics of the conductor. In layered conductors it is enough to meet the conditions $\hbar\omega_B \leqslant T$, $\hbar\omega_B \ll \varepsilon_0$, where ω_B is the cyclotron frequency of conduction electrons in a magnetic field with induction $\mathbf{B}_0$, T is temperature, $\hbar$ is Planck's constant, ε_0 is the overlap integral of the wave functions of the electrons belonging to adjacent layers.

Changing of the energy of the system for a small deviation of the distribution function δf from the equilibrium Fermi function, according to Landau– Silin Fermi liquid theory can be represented as

$$\delta E = \mathrm{Tr}_{\sigma} \int \frac{d^3\mathbf{p}' d^3\mathbf{r}}{(2\pi\hbar)^3} \hat{\varepsilon}(\mathbf{p}, \mathbf{r}, \boldsymbol{\sigma}) \delta\hat{\rho}(\mathbf{p}', \mathbf{r}, \boldsymbol{\sigma}, t)$$

$$+ \mathrm{Tr}_{\sigma'} \int \frac{d^3\mathbf{p}' d^3\mathbf{r}}{(2\pi\hbar)^3} \delta\hat{\varepsilon}(\mathbf{p}, \mathbf{r}, \boldsymbol{\sigma}, t) \delta\hat{\rho}(\mathbf{p}, \mathbf{r}, \boldsymbol{\sigma}, t). \tag{10.1}$$

The Hamilton function in one-particle approximation $\hat{\varepsilon}(\mathbf{p}, \mathbf{r}, \boldsymbol{\sigma})$, the variations of the density matrix $\delta\hat{\rho}(\mathbf{p}, \mathbf{r}, \boldsymbol{\sigma}, t)$ and the energy of quasi-particle due to electron-electron interaction

$$\delta\hat{\varepsilon}(\mathbf{p}, \mathbf{r}, t) = \mathrm{Tr}_{\sigma'} \int \frac{d^3\mathbf{p}' d^3\mathbf{r}'}{(2\pi\hbar)^3} \hat{\mathsf{F}}(\mathbf{p}, \mathbf{r}, \boldsymbol{\sigma}, \mathbf{p}', \mathbf{r}', \boldsymbol{\sigma}') \delta\hat{\rho}(\mathbf{p}', \mathbf{r}, \boldsymbol{\sigma}', t) \tag{10.2}$$

are matrices in the space of spin variables of an electron. For real metals the effective radius of electron correlations is close to the atomic distances. Therefore the case of practical interest is the one in which the correlation radius of the particles is much less than the distance over which the distribution function varies significantly, then it is possible to take

$$\hat{\mathsf{F}}(\mathbf{p}, \mathbf{r}, \boldsymbol{\sigma}, \mathbf{p}', \mathbf{r}', \boldsymbol{\sigma}') = \delta(\mathbf{r} - \mathbf{r}') \hat{\mathsf{L}}(\mathbf{p}, \boldsymbol{\sigma}, \mathbf{p}', \boldsymbol{\sigma}'). \tag{10.3}$$

The Landau correlation function $\hat{\mathsf{L}}(\mathbf{p}, \boldsymbol{\sigma}, \mathbf{p}', \boldsymbol{\sigma}')$ represents the second variational derivative of the density energy of the system with respect to distribution function of the quasi-particles. For this reason, it must be symmetrical to permutation of the variables relating to different particles. For non-magnetic system with neglect of the spin-orbit interaction, the most general form of the function $\hat{\mathsf{L}}$ is

$$\hat{\mathsf{L}}(\mathbf{p}, \hat{\boldsymbol{\sigma}}, \mathbf{p}', \hat{\boldsymbol{\sigma}}') = L(\mathbf{p}, \mathbf{p}') + S(\mathbf{p}, \mathbf{p}') \hat{\boldsymbol{\sigma}} \hat{\boldsymbol{\sigma}}'. \tag{10.4}$$

Spin-dependent operators in the right-hand part of Eq. (10.4) corresponds to the exchange interaction of the electrons.

The one-particle density matrix

$$\hat{\rho} \equiv \hat{\rho}(1) = \mathop{\mathrm{Tr}}_{2,3,\ldots N} \hat{\mathsf{R}}(1, 2, 3, \ldots N)$$

is defined as the result of convolution of the statistical operator $\hat{\mathsf{R}}(1, 2, 3, \ldots N)$ for a system of N electrons on variables of all particles except one, i.e. operation $\mathrm{Tr}_{2,3,\ldots N}$, where numbers $2, 3, \ldots N$ denote the set of dynamic variables and quantum numbers characterizing quantum states of particles.

The equation for the one-particle density matrix

$$\frac{\partial \hat{\rho}}{\partial t} + \frac{i}{\hbar}[\hat{\mathsf{H}}_1, \hat{\rho}] = \hat{I}_{coll}, \tag{10.5}$$

can be easily obtained by applying the operation $\mathrm{Tr}_{2,3,\ldots N}$ to Liouville–Neumann equation (1.1) for the statistical operator $\hat{\mathsf{R}}$. Here $\hat{\mathsf{H}}_1$ is the one-particle Hamiltonian, $[\hat{\mathsf{H}}_1, \hat{\rho}]$ is the commutator of operators $\hat{\mathsf{H}}_1$ and $\hat{\rho}$. Operator of collisions $\hat{I}_{coll}$ accurate within the normalization factor is equal to convolution Tr_2 from the commutator of the two-particle density matrix and the operator of interaction between particles. Representation of $\hat{I}_{coll}$ in the terms of the one-particle density matrix is possible only with additional simplifying assumptions, for example, in the cases of weak coupling between particles or the small density of particles.

In the coordinate representation the left side of the equation (10.5) can be written as follows:

$$\frac{\partial \hat{\rho}(\mathbf{r}_1, \mathbf{r}'_1)}{\partial t} + \frac{i}{\hbar}\int d^3\mathbf{r}''\{\hat{\mathsf{H}}_1(\mathbf{r}_1, \mathbf{r}'')\hat{\rho}(\mathbf{r}'', \mathbf{r}'_1) - \hat{\rho}(\mathbf{r}_1, \mathbf{r}'')\hat{\mathsf{H}}_1(\mathbf{r}'', \mathbf{r}'_1)\}. \tag{10.6}$$

Here $\hat{\rho}$ and $\hat{\mathsf{H}}_1$ are operators in the space of spin variables, for brevity the argument is omitted. Let us turn to the coordinate-momentum Wigner representation [22], which is defined as the Fourier transform of the coordinate representation with respect to the difference of spatial variables

$$\hat{\rho}(\mathbf{r}, \mathbf{p}) = \int d^3\mathbf{x} \exp\Big(-i\,\frac{\mathbf{px}}{\hbar}\Big)\hat{\rho}\Big(\mathbf{r} + \frac{\mathbf{x}}{2}, \mathbf{r} - \frac{\mathbf{x}}{2}\Big), \tag{10.7}$$

$$\hat{\mathsf{H}}_1(\mathbf{r}, \mathbf{p}) = \int d^3\mathbf{x} \exp\Big(-i\,\frac{\mathbf{px}}{\hbar}\Big)\hat{\mathsf{H}}_1\Big(\mathbf{r} + \frac{\mathbf{x}}{2}, \mathbf{r} - \frac{\mathbf{x}}{2}\Big), \tag{10.8}$$

here $\mathbf{x} = \mathbf{r}_1 - \mathbf{r}'_1$, $\mathbf{r} = \frac{1}{2}(\mathbf{r}_1 + \mathbf{r}'_1)$. In the spatially uniform case the quantities $\hat{\rho}(\mathbf{r}_1, \mathbf{r}'_1)$ and $\hat{\mathsf{H}}_1(\mathbf{r}_1, \mathbf{r}'_1)$ depend only on the difference $\mathbf{r}_1 - \mathbf{r}'_1$ and, therefore, Eqs. (10.7), (10.8) do not depend on the variable $\mathbf{r}$. If the characteristic size of space nonuniformity is much greater than the de Broglie wavelength λ_D of the particle, it is possible to expand the physical quantities in powers of derivatives $\partial/\partial\mathbf{r}$. This approach corresponds to transition to the quasi-classical approximation. Replacing the integration variable $\mathbf{r}''$ on $\mathbf{x}' = \mathbf{r}'' - \mathbf{r}'$ in the expression (10.6), we represent the first term in the commutator in expression (10.6) in the form

$$\int d^3\mathbf{x}'\hat{\mathsf{H}}_1\left(\mathbf{x}-\mathbf{x}',\mathbf{r}+\frac{\mathbf{x}'}{2}\right)\hat{\rho}\left(\mathbf{x}',\mathbf{r}+\frac{\mathbf{x}'-\mathbf{x}}{2}\right)$$

$$= \int d^3\mathbf{x}'\left\{\hat{\mathsf{H}}_1(\mathbf{x}-\mathbf{x}',\mathbf{r})\hat{\rho}(\mathbf{x}',\mathbf{r})+\frac{1}{2}\mathbf{x}'\frac{\partial\hat{\mathsf{H}}_1(\mathbf{x}-\mathbf{x}',\mathbf{r})}{\partial\mathbf{r}}\hat{\rho}(\mathbf{x}',\mathbf{r})\right.$$

$$\left.-\frac{1}{2}(\mathbf{x}-\mathbf{x}')\hat{\mathsf{H}}_1(\mathbf{x}-\mathbf{x}',\mathbf{r})\frac{\partial\hat{\rho}(\mathbf{x}',\mathbf{r})}{\partial\mathbf{r}}\right\}+O\left(\frac{\lambda_D^2}{L^2}\right) \quad (10.9)$$

Applying the Fourier transform to the Eq. (10.9) we obtain

$$\int d^3\mathbf{x}\exp\left(-i\frac{\mathbf{px}}{\hbar}\right)\int d\mathbf{x}'\ldots$$

$$= \hat{\mathsf{H}}_1(\mathbf{r},\mathbf{p})\hat{\rho}(\mathbf{r},\mathbf{p})+\frac{i\hbar}{2}\left(\frac{\partial\hat{\mathsf{H}}_1(\mathbf{r},\mathbf{p})}{\partial\mathbf{r}}\frac{\partial\hat{\rho}(\mathbf{r},\mathbf{p})}{\partial\mathbf{p}}-\frac{\partial\hat{\mathsf{H}}_1(\mathbf{r},\mathbf{p})}{\partial\mathbf{p}}\frac{\partial\hat{\rho}(\mathbf{r},\mathbf{p})}{\partial\mathbf{r}}\right)$$

$$+O(\hbar^2), \quad (10.10)$$

where

$$\hat{\mathsf{H}}(\mathbf{r},\mathbf{p}) = \int d^3\mathbf{x}\exp\left(-i\frac{\mathbf{px}}{\hbar}\right)\hat{\mathsf{H}}_1(\mathbf{x},\mathbf{r}), \quad \hat{\rho}(\mathbf{r},\mathbf{p}) = \int d^3\mathbf{x}\exp\left(-i\frac{\mathbf{px}}{\hbar}\right)\hat{\rho}(\mathbf{x},\mathbf{r})$$

are the main terms of the expansion of Wigner operators (10.7), (10.8) in powers of $\hbar$. Transformation of the second term in the commutator (10.6) can be obtained from formula (10.10) by means of permutation $\hat{\mathsf{H}}_1 \rightleftarrows \hat{\rho}$. In the theory of a Fermi liquid one-particle Hamiltonian $\hat{\mathsf{H}}_1(\mathbf{r},\mathbf{p})$ should be replaced by a quasi-particle energy. Then, after the Fourier transform, the relation (10.6) in the quasi-classical approximation takes the form

$$\frac{\partial\hat{\rho}}{\partial t}+\frac{i}{\hbar}[\hat{\varepsilon},\hat{\rho}]_S-\frac{1}{2}\left(\frac{\partial\hat{\varepsilon}}{\partial\mathbf{r}}\frac{\partial\hat{\rho}}{\partial\mathbf{p}}+\frac{\partial\hat{\rho}}{\partial\mathbf{p}}\frac{\partial\hat{\varepsilon}}{\partial\mathbf{r}}\right)+\frac{1}{2}\left(\frac{\partial\hat{\rho}}{\partial\mathbf{r}}\frac{\partial\hat{\varepsilon}}{\partial\mathbf{p}}+\frac{\partial\hat{\varepsilon}}{\partial\mathbf{p}}\frac{\partial\hat{\rho}}{\partial\mathbf{r}}\right) \quad (10.11)$$

where $[\hat{\varepsilon},\hat{\rho}]_S$ is the commutator of matrices in space of spin variables.

In the presence of an electromagnetic field with the vector potential $\mathbf{A}(\mathbf{r},t)$, in the expressions for $\hat{\rho}(\mathbf{r},\mathbf{p})$ and $\hat{\varepsilon}(\mathbf{r},\mathbf{p})$ it is necessary to make replacement $\mathbf{p}\to\mathbf{p}-(e/c)\mathbf{A}(\mathbf{r},t)$. The corresponding formulas of transformation for derivatives are of the form

$$\frac{\partial f}{\partial t} \to \frac{\partial f}{\partial t} - \frac{e}{c}\frac{\partial \mathbf{A}}{\partial t}\frac{\partial f}{\partial \mathbf{p}}, \quad \frac{\partial f}{\partial \mathbf{r}} \to \frac{\partial f}{\partial \mathbf{r}} - \frac{e}{c}\frac{\partial A_i}{\partial \mathbf{r}}\frac{\partial f}{\partial p_i}, \quad f = \{\hat{\rho}, \hat{\varepsilon}\}.$$

As a result, we obtain from (10.11) the following equation for the one-particle density matrix in the presence of electromagnetic field

$$\frac{\partial \hat{\rho}}{\partial t} + \frac{i}{\hbar}[\hat{\varepsilon}, \hat{\rho}]_S - \frac{1}{2}\left(\frac{\partial \hat{\varepsilon}}{\partial \mathbf{r}}\frac{\partial \hat{\rho}}{\partial \mathbf{p}} + \frac{\partial \hat{\rho}}{\partial \mathbf{p}}\frac{\partial \hat{\varepsilon}}{\partial \mathbf{r}}\right) + \frac{1}{2}\left(\frac{\partial \hat{\rho}}{\partial \mathbf{r}}\frac{\partial \hat{\varepsilon}}{\partial \mathbf{p}} + \frac{\partial \hat{\varepsilon}}{\partial \mathbf{p}}\frac{\partial \hat{\rho}}{\partial \mathbf{r}}\right)$$

$$+ e\mathbf{E}\frac{\partial \hat{\rho}}{\partial \mathbf{p}} + \frac{1}{2}\frac{e}{c}\left(\left[\frac{\partial \hat{\varepsilon}}{\partial \mathbf{p}} \times \mathbf{B}\right]\frac{\partial \hat{\rho}}{\partial \mathbf{p}} + \frac{\partial \hat{\rho}}{\partial \mathbf{p}}\left[\frac{\partial \hat{\varepsilon}}{\partial \mathbf{p}} \times \mathbf{B}\right]\right) = \hat{I}_{coll}(\hat{\rho}), \quad (10.12)$$

where $\hat{I}_{coll}$ – the operator of collisions transformed by means of Fourier transform, $\mathbf{E}$ – electric field, $\mathbf{B} = \mathbf{B}_0 + \mathbf{B}^{\sim}(\mathbf{r}, t)$, $\mathbf{B}^{\sim}(\mathbf{r},t)$ – induction of the alternating magnetic field. Operator

$$\hat{\varepsilon} = \varepsilon(\mathbf{p})\delta_{\alpha\beta} - \mu_0 \boldsymbol{\sigma}\mathbf{B} + \delta\hat{\varepsilon}(\mathbf{p}, \mathbf{r}, t) \quad (10.13)$$

is a sum of the energy of a quasiparticle in one-electron approximation in the magnetic field and the energy of the quasiparticle $\delta\hat{\varepsilon}(\mathbf{p}, \mathbf{r}, t)$ due to interelectron interaction, μ_0 – magnetic moment of an electron, $\delta_{\alpha\beta}$ – Kronecker delta.

The alternating electric $\mathbf{E}$ and magnetic $\mathbf{B}^{\sim}(\mathbf{r},t)$ fields are defined from the Maxwell equations

$$\mathrm{rot}\mathbf{B}^{\sim} = \frac{1}{c}\frac{\partial \mathbf{E}}{\partial t} + \frac{4\pi}{c}\mathbf{J}, \quad \mathrm{rot}\mathbf{E} = -\frac{1}{c}\frac{\partial \mathbf{B}^{\sim}}{\partial t}, \quad \mathrm{div}\mathbf{B}^{\sim} = 0, \quad (10.14)$$

supplemented with the material equation for the current density induced in the medium

$$\mathbf{J}(\mathbf{r}, t) = e\mathrm{Tr}_\sigma \int \frac{d^3\mathbf{p}}{(2\pi\hbar)^3}\hat{\rho}(\mathbf{p}, \mathbf{r}, \boldsymbol{\sigma}, t)\frac{\partial \hat{\varepsilon}}{\partial \mathbf{p}} + c\mu_0 \,\mathrm{rot}\, \mathrm{Tr}_\sigma \int \frac{d^3\mathbf{p}}{(2\pi\hbar)^3}\boldsymbol{\sigma}\hat{\rho}(\mathbf{p}, \mathbf{r}, \boldsymbol{\sigma}, t). \quad (10.15)$$

Instead of the matrix equation (10.12), it is convenient to consider a system of four equations for the distribution function

$$f(\mathbf{r}, \mathbf{p}, t) = \mathrm{Tr}_\sigma \hat{\rho}$$

and the spin density

$$\mathbf{g}(\mathbf{r}, \mathbf{p}, t) = \mathrm{Tr}_\sigma(\boldsymbol{\sigma}\hat{\rho})$$

One of these equations is obtained by applying the operation of taking the trace with respect to the spin variables to the matrix equation for $\hat{\rho}$, while the other three are obtained by applying the operation Tr to the original equation (10.12) multiplied by $\boldsymbol{\sigma}$. The function $\mathbf{g}(\mathbf{r}, \mathbf{p}, t)$, together with the second term on the right-hand side of (10.15), describes the spin waves.

Representing the collision integral in the form

$$\hat{I}_{coll} = \hat{I}^{(1)}_{coll}\delta_{\alpha\beta} + \hat{\mathbf{I}}^{(2)}_{coll}\boldsymbol{\sigma}, \tag{10.16}$$

after simple transformations we obtain

$$\frac{\partial f}{\partial t} + \{f, \varepsilon_1\} + \{\boldsymbol{\varepsilon}_2, \mathbf{g}\} + e\mathbf{E}\frac{\partial f}{\partial \mathbf{p}} + \frac{e}{c}\left(\left[\frac{\partial \varepsilon_1}{\partial \mathbf{p}} \times \mathbf{B}\right]\frac{\partial f}{\partial \mathbf{p}} + \left[\frac{\partial \varepsilon_{2i}}{\partial \mathbf{p}} \times \mathbf{B}\right]\frac{\partial g_i}{\partial \mathbf{p}}\right) = \hat{I}^{(1)}_{coll}(f),$$

$$\begin{aligned}\frac{\partial \mathbf{g}}{\partial t} &+ \left(\frac{\partial \varepsilon_1}{\partial \mathbf{p}}\frac{\partial}{\partial \mathbf{r}}\right)\mathbf{g} - \left(\frac{\partial \varepsilon_1}{\partial \mathbf{r}}\frac{\partial}{\partial \mathbf{p}}\right)\mathbf{g} + \frac{2}{\hbar}[\boldsymbol{\varepsilon}_2 \times \mathbf{g}] \\ &+ \left(\frac{\partial f}{\partial \mathbf{r}}\frac{\partial}{\partial \mathbf{p}}\right)\boldsymbol{\varepsilon}_2 - \left(\frac{\partial f}{\partial \mathbf{p}}\frac{\partial}{\partial \mathbf{r}}\right)\boldsymbol{\varepsilon}_2 + e\left(\mathbf{E}\frac{\partial}{\partial \mathbf{p}}\right)\mathbf{g} \\ &+ \frac{e}{c}\left(\left[\frac{\partial \varepsilon_1}{\partial \mathbf{p}} \times \mathbf{B}\right]\frac{\partial}{\partial \mathbf{p}}\right)\mathbf{g} - \frac{e}{c}\left(\left[\frac{\partial f}{\partial \mathbf{p}} \times \mathbf{B}\right]\frac{\partial}{\partial \mathbf{p}}\right)\boldsymbol{\varepsilon}_2 = \hat{\mathbf{I}}^{(2)}_{coll}(\mathbf{g}),\end{aligned} \tag{10.17}$$

here

$$\{\mathbf{g}, \boldsymbol{\varepsilon}_2\} = \frac{\partial \mathbf{g}}{\partial r_i}\frac{\partial \boldsymbol{\varepsilon}_2}{\partial p_i} - \frac{\partial \boldsymbol{\varepsilon}_2}{\partial r_i}\frac{\partial \mathbf{g}}{\partial p_i}, \quad \varepsilon_1 = \varepsilon(\mathbf{p}) + \int \frac{d^3\mathbf{p}'}{(2\pi\hbar)^3} L(\mathbf{p}, \mathbf{p}')\delta f(\mathbf{p}', \mathbf{r}, t),$$

$$\boldsymbol{\varepsilon}_2 = -\mu_0\mathbf{B} + \int \frac{d^3\mathbf{p}'}{(2\pi\hbar)^3} S(\mathbf{p}, \mathbf{p}')\delta\mathbf{g}(\mathbf{p}', \mathbf{r}, t).$$

The system of equations (10.14), (10.15), (10.17) describes the eigen oscillations of the electromagnetic field and the spin density in the conductors with arbitrary energy spectrum and correlation function of Landau.

We use the equations (10.17) for calculation of collective processes in layered conductors with electron with the energy spectrum (2.1)

$$\varepsilon(\mathbf{p}) = \varepsilon_0(p_x, p_y) + \sum_{n=1}^{\infty} \varepsilon_n(p_x, p_y, \eta)\cos\frac{np_z}{p_0}$$

where functions $\varepsilon_n(p_x, p_y, \eta)$ significantly decrease with growth of their number and the greatest of them $\varepsilon_1(p_x, p_y, \eta) \simeq \eta\varepsilon_F$.

For angles ϑ between $\mathbf{B}_0 = (B_0 \sin\vartheta, 0, B_0 \cos\vartheta)$ and $\mathbf{n}$ that are not too close to $\pi/2$, namely, for $\pi/2 - \vartheta \gg \eta$, closed electron orbits in the momentum space for different values of the momentum projection onto the magnetic field direction are nearly indistinguishable, while the area $S(\varepsilon, p_B)$ of the cross section of the FS by the plane $p_B = \mathbf{p}\mathbf{B}_0/B = p_z \cos\vartheta + p_x \sin\vartheta = const$, and the components v_x and v_y of the velocity $\mathbf{v} = \partial\varepsilon(\mathbf{p})/\partial\mathbf{p}$ of conduction electrons in the plane of layers depend weakly on p_B. This means that the energy of quasiparticles in the one-electron approximation, the Landau correlation function, and the cyclotron frequency can be expanded into an asymptotic series in the quasi-two-dimensionality parameter η, and the main term of the asymptotics is independent of p_B. In the zeroth-order approximation in η, the functions $L(\mathbf{p}, \mathbf{p}')$ and $S(\mathbf{p}, \mathbf{p}')$ can be represented as the Fourier series

$$L(\mathbf{p}, \mathbf{p}') = \sum_{n=-\infty}^{\infty} L_n(\varepsilon_F) e^{in(\varphi-\varphi')}, \quad S(\mathbf{p}, \mathbf{p}') = \sum_{n=-\infty}^{\infty} S_n(\varepsilon_F) e^{in(\varphi-\varphi')}. \tag{10.18}$$

Formulas (10.18) are analogs of series expansion on Legendre's polynomials of correlation function of quasiparticles with isotropic energy spectrum. The integrals of motion of charge carriers in a magnetic field, ε and p_B, as well as the phase of the electron velocity $\varphi = \omega_B t$, where t is the time of motion along a trajectory $\varepsilon = \varepsilon_F$, p_B, are chosen as the variables in the $\mathbf{p}$-space. Due to the symmetry $\mathsf{L}(\mathbf{p}, \hat{\boldsymbol{\sigma}}, \mathbf{p}', \hat{\boldsymbol{\sigma}}') = \mathsf{L}(\mathbf{p}', \hat{\boldsymbol{\sigma}}', \mathbf{p}, \hat{\boldsymbol{\sigma}})$ with respect to the permutation of arguments, the coefficients in (10.18) are related by the formulas $L_{-n} = L_n$, $S_{-n} = S_n$. Consideration of the subsequent terms of the expansion of the correlation function in powers of η results in negligibly small corrections to kinetic coefficients.

For small deviations of the electron system from equilibrium, we can represent the functions f and $\mathbf{g}$ as the respective sums of equilibrium parts and small non-equilibrium components,

$$\begin{aligned} f(\mathbf{r}, \mathbf{p}, t) &= f_0(\varepsilon + \delta\varepsilon) - \psi(\mathbf{r}, \mathbf{p}, t)\frac{\partial f_0}{\partial \varepsilon}, \\ \mathbf{g}(\mathbf{r}, \mathbf{p}, t) &= \mathbf{g}_0(\varepsilon) - \frac{\partial f_0}{\partial \varepsilon}\, \boldsymbol{\xi}(\mathbf{r}, \mathbf{p}, t), \end{aligned} \tag{10.19}$$

Here, $f_0(\varepsilon)$ is the Fermi function, and $\mathbf{g}_0(\varepsilon) = -\mu\mathbf{B}_0(\partial f_0/\partial\varepsilon)$, $\mu = \mu_0/(1 + \nu(\varepsilon_F)S_0)$. The integral of $\mu_0\mathbf{g_0}(\varepsilon) = -\mu\mu\mathbf{B}_0(\partial f_0/\partial\varepsilon)$ over a unit cell in the **p**-space represents the magnetization $\mathbf{M}_0 = \chi_0\mathbf{B}_0$ in a uniform constant magnetic field of induction $\mathbf{B}_0$, $\chi_0 = \mu_0\mu\nu(\varepsilon_F)$ is the static paramagnetic susceptibility, and $\nu(\varepsilon_F)$ is the density of states at the Fermi level. The non-equilibrium component of the distribution function satisfies the linearized Boltzmann equation

$$\frac{\partial\psi}{\partial t} + \left(\mathbf{v}\frac{\partial}{\partial\mathbf{r}} + \frac{e}{c}(\mathbf{v}\times\mathbf{B}_0)\frac{\partial}{\partial\mathbf{p}}\right)(\psi + \langle L\psi\rangle) + e\mathbf{v}\mathbf{E} = \hat{I}^{(1)}_{coll}(\psi + \langle L\psi\rangle), \tag{10.20}$$

while the kinetic equation for the perturbed spin density in the case when ξ is perpendicular to $\mathbf{B}_0$ is given, according to [132, 133], by

$$\frac{\partial\xi}{\partial t} + \left(\mathbf{v}\frac{\partial}{\partial\mathbf{r}} + \frac{e}{c}(\mathbf{v}\times\mathbf{B}_0)\frac{\partial}{\partial\mathbf{p}}\right)(\xi + \langle S\xi\rangle) - \frac{2\mu}{\hbar}[\mathbf{B}_0\times(\xi + \langle S\xi\rangle)]$$
$$-\mu_0\mathbf{v}\frac{\partial\mathbf{B}^{\sim}}{\partial\mathbf{r}} + \frac{2\mu\mu_0}{\hbar}[\mathbf{B}_0\times\mathbf{B}^{\sim}] = \hat{I}^{(2)}_{coll}(\xi + \langle S\xi\rangle), \tag{10.21}$$

where angular brackets denote averaging over the Fermi surface:

$$\langle L\psi\rangle \equiv \int\frac{2d^3\mathbf{p}'}{(2\pi\hbar)^3}\left(-\frac{\partial f_0(\varepsilon')}{\partial\varepsilon'}\right)L(\mathbf{p},\mathbf{p}')\psi(\mathbf{r},\mathbf{p}',t).$$

Current density $\mathbf{J} = \mathbf{j} + \mathbf{j}^{(\mathbf{m})}$, consisting of conductivity current $\mathbf{j}$ and magnetization current $\mathbf{j}^{(\mathbf{m})}$, is expressed in terms of functions $\Psi \equiv \psi + \langle L\psi\rangle$ and ξ, making use of expression (10.15) we have

$$\mathbf{j} = e\langle\mathbf{v}(\psi + \langle L\psi\rangle)\rangle, \tag{10.22}$$

$$\mathbf{j}^{(\mathbf{m})} = c\mu_0\,\mathrm{rot}\langle\xi\rangle. \tag{10.23}$$

Let us find the equations that determine distribution function $\Psi \equiv \psi + \langle L\psi\rangle$ and spin density $\mathbf{\Phi} \equiv \xi + \langle S\xi\rangle$ renormalized by Fermi-liquid interaction. Expanding the functions ψ and ξ into Fourier series in the variable φ,

$$\psi = \sum_{n=-\infty}^{\infty}\psi_n(\varepsilon, p_B)e^{in\varphi}, \quad \xi = \sum_{n=-\infty}^{\infty}\xi_n(\varepsilon, p_B)e^{in\varphi}.$$

and applying formulas (10.18), we obtain:

$$\psi = \Psi - \sum_{p=-\infty}^{\infty} \kappa_p \bar{\Psi}_p e^{ip\varphi}, \tag{10.24}$$

$$\xi = \mathbf{\Phi} - \sum_{p=-\infty}^{\infty} \lambda_p \bar{\mathbf{\Phi}}_p e^{ip\varphi}, \tag{10.25}$$

where

$$\bar{\mathbf{\Phi}}_p = \frac{1}{(2\pi)^2} \int_0^{2\pi} d\varphi \int_{-\pi}^{\pi} d\beta e^{-ip\varphi} \mathbf{\Phi}(\varepsilon_F, \beta, \varphi) \equiv \langle e^{-ip\varphi} \mathbf{\Phi} \rangle_{\beta,\varphi}, \tag{10.26}$$

$$\kappa_p = \frac{L_p^{\sim}}{1 + L_p^{\sim}}, \lambda_p = \frac{S_p^{\sim}}{1 + S_p^{\sim}}, \beta = \frac{p_B}{p_0 \cos\vartheta}, \; S_n^{\sim} = \nu(\varepsilon_F) S_n, L_p^{\sim} = \nu(\varepsilon_F) L_p.$$

Assuming that the space-time dependence of the variable quantities is given by $\exp(-i\omega t + i\mathbf{kr})$ and substituting the relation (10.24) into the Eq. (10.20) we obtain the following equation of the Fourier components of the function Ψ.

$$\frac{\partial \Psi}{\partial \varphi} - \frac{i}{\omega_B}(\omega - \mathbf{kv})\Psi = \frac{e\mathbf{vE}}{\omega_B} - \frac{i\omega}{\omega_B} \sum_{p=-\infty}^{\infty} \kappa_p \bar{\Psi}_p(\varepsilon) e^{ip\varphi} + \frac{1}{\omega_B} I_{coll}^{(1)}(\Psi). \tag{10.27}$$

Using Eqs. (10.25) and (10.21) we find that the Fourier components of the renormalized spin density $\Phi^{(\pm)} = \Phi_{x_1} \pm i\Phi_y$ of conduction electrons with the quasi-two-dimensional dispersion law satisfy the equation

$$\frac{\partial \Phi_\pm}{\partial \varphi} - \frac{i}{\omega_B}(\omega - \mathbf{kv} \mp \Omega_s)\Phi_\pm$$
$$= i\frac{\mu_0}{\omega_B}(\mathbf{kv} \pm \Omega_s)\mathbf{B}_\pm^{\sim} - \frac{i\omega}{\omega_B} \sum_{n=-\infty}^{\infty} \lambda_n \bar{\Phi}_n^{(\pm)} e^{in\varphi} + \frac{1}{\omega_B} I_{coll}^{(2)}(\Phi_\pm) \tag{10.28}$$

Here $\Phi_{x_1} = \Phi_x \cos\vartheta - \Phi_z \sin\vartheta$, the axis x_1 is perpendicular to the axis y and the vector $\mathbf{B}_0$, $B_\pm^{\sim} = B_x \pm iB_y$

$$\Omega_s = \frac{\omega_s}{1 + S_0^{\sim}} \tag{10.29}$$

and $\omega_s = -2\mu_0 B_0/\hbar$ is the spin paramagnetic resonance frequency.

The collision integrals $\hat{I}^{(1)}_{coll}$ and $\hat{I}^{(2)}_{coll}$ determine the characteristic relaxation times of momentum τ_1 and spin density τ_2, usually $\tau_2 \gg \tau_1$. In the equation (10.28) we use τ-approximation

$$\hat{I}^{(1)}_{coll}(\Psi) = -\frac{1}{\tau_1}\Psi, \qquad (10.30)$$

i.e. we replace the linear operator $\hat{I}^{(1)}_{coll}$ by operator of multiplication on the inverse relaxation time. For the collision integral $\hat{I}^{(2)}_{coll}$ it is possible to take the following model expression

$$\hat{I}^{(2)}_{coll}(\Phi) = -\frac{1}{\tau_2}\Phi - \frac{1}{\tau_1}(\Phi - \bar{\Phi}_0), \qquad (10.31)$$

considering both spin and orbital mechanisms of the relaxation. Henceforth, we will consider processes that correspond to the range of frequencies $\omega \gg \tau^{-1} = \tau_1^{-1} + \tau_2^{-1}$, where the asymptotics of the spectrum of collective modes is independent of the specific form of the collision integral.

Let us integrate the equations (10.27) and (10.28), assuming formally that quantities $\bar{\Phi}_n$ and $\bar{\Psi}_n$ are known. As a result we obtain the integral equations for renormalized distribution function and spin density [132,133]

$$\Psi = \int_{-\infty}^{\varphi} d\varphi' \exp\left(-\frac{i}{\omega_B}\int_{\varphi'}^{\varphi} d\varphi''\,(\omega + i/\tau_1 \mp \omega_B - \mathbf{kv}(\varphi'', \beta))\right) \times \left(\frac{e\mathbf{vE}}{\omega_B} - i\frac{\omega}{\omega_B}\sum_{p=-\infty}^{\infty} \kappa_p \bar{\Psi}_p e^{ip\varphi'}\right) \qquad (10.32)$$

$$\Phi^{(\pm)} = \int_{-\infty}^{\varphi} d\varphi' \exp\left(-\frac{i}{\omega_B}\int_{\varphi'}^{\varphi} d\varphi''\,(\omega + i/\tau \mp \Omega_s - \mathbf{kv}(\varphi'', \beta))\right) \times \left(i\frac{\mu_0}{\omega_B}(\mathbf{kv} \pm \Omega_s)\mathbf{B}^{\sim}_{\pm} - i\frac{\omega}{\omega_B}\sum_{p=-\infty}^{\infty} \lambda_p \bar{\Phi}^{(\pm)}_p e^{ip\varphi'} + \frac{1}{\omega_B\tau_1}\bar{\Phi}^{(\pm)}_0\right), \qquad (10.33)$$

that do not explicitly contain the functions ψ and ξ. On the basis of the equations (10.32) and (10.33) we can calculate the kinetic coefficients of the electron Fermi-liquid of Q2D conductors.

11. Fermi-liquid modes

The specific features of quasi-two-dimensional energy spectrum in layered conductors give rise to peculiar Fermi-liquid modes, which are absent in the gas of charge carriers [124]. These collective excitations with frequencies much lower than the plasma frequency can propagate along the normal to the layers in the presence of a strong magnetic field. Due to the correlation effects there are transparent windows for electromagnetic waves with different polarizations even at weak intensity of Fermi-liquid interaction.

Let us use the approximation (7.3) for the energy spectrum of electrons. In the case when the wave vector $\mathbf{k} = (0, 0, k)$ is aligned parallel to direction of magnetic field $\mathbf{B}_0 = (0, 0, B_0)$, the dispersion equation (8.1), which determins the spectrum $\omega(\mathbf{k})$ of the eigenmodes of the electromagnetic field, splits into three equations

$$\varepsilon_{zz} = 0, \quad \varepsilon_{xx} \pm i\varepsilon_{xy} = \left(\frac{kc}{\omega}\right)^2, \tag{11.1}$$

(we have taken into account the equations $\varepsilon_{xx} = \varepsilon_{yy}$, $\varepsilon_{xz} = \varepsilon_{yz} = 0$ because of the isotropy in the layer plane of the electron dispersion law). The first of these equations describes purely longitudinal oscillations and coincides exactly with the dispersion equation for longitudinal oscillations of the charged Fermi liquid in the absence of a magnetic field. The other two equations describe transverse electromagnetic waves with different polarizations. Circularly polarized components of the electric field $E_\pm = E_x \pm iE_y$ of these waves and of the electric current density are connected by the simple relationship

$$j_\pm = j_x \pm ij_y = (\sigma_{xx} \mp i\sigma_{xy})E_\pm. \tag{11.2}$$

According to Eqs. (7.4) and (7.6), the asymptotic expressions for the electron velocity in the plane of the layers $v_\perp$ and the cyclotron frequency ω_B accurate to the terms of the order of η are equal v_F and $|e|B_0/mc$, respectively. Using the expression (10.22) and Eqs. (7.4), we find

$$j_\pm = e\nu(\varepsilon_F)\langle(v_x \pm iv_y)\Psi\rangle_{\beta,\varphi} = ev_F\nu(\varepsilon_F)\bar{\Psi}_{\pm 1}, \tag{11.3}$$

where $\bar{\Psi}_{\pm 1} = \langle e^{\pm i\varphi}\Psi\rangle_{\beta,\varphi}$ is determined by the formula (10.26) and $\nu(\varepsilon_F) = mp_0/\pi\hbar^3$ is the density of states of electrons with the dispersion law (7.3). The coefficients $\bar{\Psi}_{\pm 1}$ can be readily found from Eq. (10.32), being transformed to the form

$$\Psi = \frac{i}{2}\frac{ev_\perp E_+ e^{i\varphi}}{\tilde{\omega} - kv_z - \omega_B} + \frac{i}{2}\frac{ev_\perp E_- e^{-i\varphi}}{\tilde{\omega} - kv_z + \omega_B} + \omega \sum_{n=-\infty}^{\infty} \kappa_n \frac{\bar{\Psi}_n e^{in\varphi}}{\tilde{\omega} - kv_z - n\omega_B}, \tag{11.4}$$

where $\tilde{\omega} = \omega + i/\tau_1$. Multiplying Eq. (11.4) on $e^{\pm i\varphi}$ and integrating the result over $d\varphi$ and $d\beta$ we obtain the algebraic equation for $\bar{\Psi}_{\pm 1}$

$$\bar{\Psi}_{\pm 1} = (i\pi e v_\perp E_\pm + \omega\kappa_1 \bar{\Psi}_{\pm 1})\left\langle \frac{1}{\tilde{\omega} - kv_z \pm \omega_B} \right\rangle_\beta. \tag{11.5}$$

With the aid of Eqs. (11.2), (11.3) and (11.5), it is easy to find $\sigma_{xx} \mp i\sigma_{xy}$ and represent the dispersion equation for transverse waves in the collisionless limit ($\tau_1 \to \infty$) as

$$\left(\frac{kc}{\omega}\right)^2 = 1 - \frac{\omega_p^2}{\omega}\left(\operatorname{sign}(\omega \mp \omega_B)\sqrt{(\omega \mp \omega_B)^2 - (\eta k v_F)^2} - \kappa_1\omega\right)^{-1}. \tag{11.6}$$

For the frequencies much lower than the plasma frequency $\omega_p = (4\pi e^2 n_0/m)^{1/2}$, ($n_0$ is the density of charge carriers), this equation at $\kappa_1 > 0$ has real solutions

$$\omega^{(\pm)} = \frac{\sqrt{(\eta k v_F)^2 + [\omega_B^2 - (\eta k v_F)^2](\kappa_1 - \omega_p^2/k^2c^2)^2} \pm \omega_B}{1 - (\kappa_1 - \omega_p^2/k^2c^2)^2} \tag{11.7}$$

defining two branches of eigenmodes of electromagnetic field. The low-frequency branch $\omega^{(-)} < \omega_B$ in the limiting case when $\eta k v_F$, $\omega \ll \omega_B$ represents a helicoidal wave with frequency $\omega = \omega_B k^2 c^2/\omega_p^2$, which propagates along an external magnetic field direction. The other, a high-frequency wave $\omega^{(+)} > \omega_B$, results from the Fermi-liquid interaction between charge carriers and exists if the following conditions hold:

$$(\omega \mp \omega_B)^2 - (\kappa_1\omega)^2 < (\eta k v_F)^2 < (\omega \mp \omega_B)^2, \quad k > \frac{\omega_p}{c\sqrt{\kappa_1}}. \tag{11.8}$$

The inequalities (11.8) show that the frequency takes real values at real values of the wave vector. The inequality $(\eta k v_F)^2 < (\omega \mp \omega_B)^2$ is a usual condition of the absence of Landau collisionless attenuation in a magnetic field. When this condition is satisfied the attenuation rate of the wave is determined by collision processes and is proportional to τ_1^{-1}. The values $k_{\min} = \omega_p/c\sqrt{\kappa_1}$ and $\omega^{\pm}(k_{\min}) = \eta\omega_p v_F/c\sqrt{\kappa_1} \pm \omega_B$ correspond to the edge of the wave spectrum. The dependence of limiting values for the wave vector and the frequency on κ_1 shows that the waves under consideration can exist at however weak a Fermi-liquid interaction. It suffices that $\kappa_1 > 0$, but the conditions for the excitation of these modes become more difficult with decreasing of κ_1.

Proceeding to the limit $\omega_B \to 0$ in Eq. (11.7), we obtain the spectrum of the Fermi-liquid wave in the absence of a magnetic field,

$$\omega = \frac{\eta k v_F}{\sqrt{1 - (\kappa_1 - \omega_p^2/k^2c^2)^2}}. \tag{11.9}$$

The magnetic field lifts the degeneracy from the spectrum, which leads to the appearance of two waves with different polarizations, the limiting value for the frequency $\omega^{(-)}$ being decreased. It is easily seen from Eq. (11.7) that at $\mathbf{B}_0 \to 0$ the value of $k_{\min}$ remains unchanged.

12. Spin waves in strongly anisotropic organic conductors

In sections 8 and 9, we have considered electromagnetic waves occurring under the resonance conditions, which are caused by the cyclotron and the orbital motion of charge carriers. Oscillations of a quite different type are paramagnetic spin waves conditioned by the presence of intrinsic magnetic moment of an electron and associated with the spin paramagnetic resonance. Spin waves cannot exist in gas plasma or other plasma medium in which Fermi-liquid interaction is negligible. A necessary condition of their excitation is the existence of rather strong exchange interaction between electrons. At the propagation of spin collective modes, the new mechanism of collisionless damping is added to those considered in section 8. The mechanism is attributed to the spin resonance, i.e. the absorption of electromagnetic energy under magnetic dipole transitions between Zeeman electron energy levels in a magnetic field. These transitions are caused by an alternating magnetic field polarized perpendicular to $\mathbf{B}_0$ with a frequency close to Ω_s. The conditions of maxim absorption for spin modes similar to (7.1) has the form

$$\omega - n\omega_B \mp \Omega_s - \mathbf{kv} = 0. \tag{12.1}$$

Signs $\mp$ correspond to different directions of rotation of the magnetic field of a wave in a plane perpendicular to vector $\mathbf{B}_0$ and Ω_s is defined by formula (10.29). In Q2D conductors the quasi-classical description of kinetic processes is valid when the overlap integral of wave functions of electrons belonging to adjacent layers is significantly greater than the characteristic frequencies of collective modes, i.e.,

$$\varepsilon_0 \gg \hbar\omega_B, \hbar\Omega_s.$$

Properties of spin waves are determined by the magnetic susceptibility tensor $\chi_{ij}(\omega, \mathbf{k})$, which can be written as the product of the static paramagnetic susceptibility χ_0 and a tensor function that determines the dispersion properties of the medium. In non-magnetic metals, the value of χ_0 is of the order of 10^{-6}. Due to the smallness of the effect of paramagnetism in solid-state plasma its high-frequency properties can manifest only under the resonance conditions, when $\max|\chi_{ij}(\omega, \mathbf{k})| \gg \chi_0$. If the spin-orbit interaction can be neglected, the spin density oscillations are not connected with oscillations of quasiparticle distribution function; as a result spin waves and electromagnetic waves can be considered independently.

Experimental observation of spin waves is very difficult because of the small value of the paramagnetic susceptibility. Spin waves were observed in alkali metals when studying the selective transparency of thin metal films [131]. High-frequency electromagnetic field induces the non-equilibrium magnetization in the skin layer. Since the spin relaxation time (spin-flip time) is much greater than the momentum relaxation time $\tau_2 \gg \tau_1$, electrons can transfer the magnetization inside the metal at a distance significantly greater than the skin depth, even at $\omega_B\tau_1 \ll 1$. Under this condition, the magnetization dynamics is diffusive, i.e. magnetization can be described by diffusion equation [114]. If the film thickness is less than the distance traveled by the electron during the spin flip, there is transparency of the film at a frequency equal to the frequency of spin paramagnetic resonance ω_s. At low temperatures, in pure samples, when the momentum relaxation time is sufficiently large $\omega_B\tau_1 \gg 1$, the magnetization can be propagated in the metal in the form of spin waves. As a result, additional transparency lines appear in thin films corresponding to the frequencies of the spin waves. EPR was observed in organic metals of the family BEDT-TTF salts, see Fig. 11, [100–102]. Under experimental conditions, the movement of electrons inside the sample occurred along the normal to the layers, with the velocity η times less than the characteristic velocity of the electrons in the plane of the layers. For this reason, the diffuse mechanism of the occurrence of ESR apparently is ineffective. Most probably, that EPR absorption peaks in [100–102] are due to the excitation of the spin collective modes.

Propagation of spin waves in layered conductors is characterized by the same specific features as the propagation of electromagnetic collective modes which are common to all elementary excitations of the Bose type. The drift velocity of the charge carriers is an oscillating function of the angle between the magnetic field and the direction of lowest conductivity, for certain orientations of the magnetic field, it is a negligible value. For these directions of the magnetic field there is no Landau damping and weakly damped waves may propagate even under strong spatial dispersion. In the short-wavelength limit collective modes may exist with frequencies in the vicinity of resonances at any orientation of wave vector with respect to the magnetic field. Similar types of excitations in quasi-isotropic metals are possible only in the direction perpendicular to the direction of the external magnetic field.

The dispersion equation for spin waves

Let us use the Eq. (10.33) for renormalized spin density in the collisionless limit. Multiplying this equation by $e^{-in\varphi}$ and integrating the result with respect to the variables $\beta = p_B/p_0 \cos\vartheta$ and φ, for the Fourier coefficients $\bar{\Phi}_n^{(\pm)}$ of the function

$$\langle \Phi^{(\pm)} \rangle_\beta = \frac{1}{2\pi} \int_0^{2\pi} d\beta \Phi^{(\pm)}(\varepsilon_F, \beta, \varphi)$$

we obtain the following system of linear equations

$$\sum_{p=-\infty}^{\infty} \left(\delta_{np} - \lambda_p \frac{\omega}{\omega_B} \langle g_{np}(\beta) \rangle_\beta \right) \bar{\Phi}_p^{(\pm)} = -\mu_0 B_{\pm}^{\sim} \langle F_n \rangle_\beta. \qquad (12.2)$$

Here

$$g_{np}(\beta) = \frac{1}{2\pi i}\left(1 - \exp\frac{2\pi i}{\omega_B}(\tilde{\omega} \mp \Omega_s - \langle \mathbf{kv}\rangle_\varphi)\right)^{-1} \int_0^{2\pi}\int_0^{2\pi} d\varphi d\varphi_1$$

$$\times \exp\left(i(p-n)\varphi - ip\varphi_1 + \frac{\tilde{\omega} \mp \Omega_s}{\omega_B}\varphi_1 - \frac{i}{\omega_B}\int_{\varphi-\varphi_1}^{\varphi} d\varphi' \mathbf{kv}(\varphi')\right)$$

(12.3)

$$F_n = \frac{1}{2\pi i \omega_B}\left(1 - \exp\frac{2\pi i}{\omega_B}(\tilde{\omega} \mp \Omega_s - \langle \mathbf{kv}\rangle_\varphi)\right)^{-1}$$

$$\times \int_0^{2\pi}\int_0^{2\pi} d\varphi d\varphi_1 (\mathbf{kv}(\varphi - \varphi_1) \mp \Omega_s)$$

$$\times \exp\left(i(p-n)\varphi - ip\varphi_1 + i\frac{\tilde{\omega} \mp \Omega_s}{\omega_B}\varphi_1 - \frac{i}{\omega_B}\int_{\varphi-\varphi_1}^{\varphi} d\varphi' \mathbf{kv}(\varphi')\right)$$

The alternating magnetic field, induced by spin oscillations, is determined by the equation

$$\mathbf{B}^{\sim}(\omega, \mathbf{k}) = 4\pi\left(\mathbf{M}^{\sim}(\omega, \mathbf{k}) - \frac{\mathbf{k}}{k^2}(\mathbf{k}\mathbf{M}^{\sim}(\omega, \mathbf{k}))\right), \qquad (12.4)$$

where $\mathbf{M}^{\sim}(\omega, \mathbf{k}) = \mu_0\langle \xi(\mathbf{p}, \omega, \mathbf{k})\rangle = \mu\nu(\varepsilon_F)\bar{\mathbf{\Phi}}_0(\omega, \mathbf{k})$ is the high-frequency magnetization.

The Fourier coefficients of the smooth function $\nu(\varepsilon_F)S(\mathbf{p}, \mathbf{p}')$ rapidly decrease as their number increases; therefore, we may restrict oneself to a finite number of terms of the series. The system of equations (12.2), combined with Eq. (12.4), which links the high-frequency magnetic field to the magnetization, describes natural oscillations of spin density in layered conductors with arbitrary energy spectrum and correlation function. Using Eq. (12.2), it is easy to obtain the magnetic susceptibility, taking into account time and spatial dispersion

$$\chi_{\pm}(\omega,\mathbf{k}) = \mu\nu(\varepsilon_F)\frac{\bar{\Phi}_0^{(\pm)}(\omega,\mathbf{k})}{B_{\pm}^{\sim}}$$

$$= \chi_0 \frac{\det\left[\delta_{0p}\langle F_n\rangle_\beta + \left(\delta_{np} - \lambda_p\frac{\omega}{\omega_B}\langle g_{np}(\beta)\rangle_\beta\right)(1-\delta_{0p})\right]}{\det\left[\delta_{np} - \lambda_p\frac{\omega}{\omega_B}\langle g_{np}(\beta)\rangle_\beta\right]} \quad (12.5)$$

For the frequencies, that do not coincide with the frequency of eigen-oscillations of the spin density, the components $\chi_{ik}(\omega,\mathbf{k})$ are of the order of the static paramagnetic susceptibility $\chi_0 \simeq \mu_0^2\nu(\varepsilon_F) \sim 10^{-6}$. For this reason, in order to find the spectrum of the spin waves, it is sufficient to use the homogeneous system of equations corresponding to Eq. (12.2). In the expression (12.2) we can neglect the small non-uniform correction proportional to $\mu_0 B_{\pm}^{\sim}$ which allows for the influence of the self-coordinated field $B_{\pm}^{\sim}$. The dispersion equation for «free» oscillations of the spin density is of the form

$$D(\omega^{(0)},\mathbf{k}) \equiv \det\left[\delta_{np} - \lambda_p\frac{\omega^{(0)}}{\omega_B}\langle g_{np}(\beta)\rangle_\beta\right] = 0. \quad (12.6)$$

The frequency of eigen-oscillations of the magnetization within the terms proportional to $\chi_0 \simeq \mu_0^2\nu(\varepsilon_F) \simeq 10^{-6}$, coincides with the frequency of the spin density «free» oscillations. At this frequency the magnetic susceptibility has a sharp maximum, and the determinant $D(\omega,\mathbf{k})$ is of the order of χ_0.

The condition of the absence of collisionless attenuation of spin waves is similar to the condition (8.5) for electromagnetic collective modes and it is also reduced to the performance of the inequality

$$|\omega - n\omega_B \mp \Omega_s| > \max|\langle\mathbf{kv}\rangle_\varphi|. \quad (12.7)$$

In the opposite case the integrand in the formula (12.3) has a pole and after integration with respect to p_B the dispersion equation (12.6) acquires an imaginary part responsible for the strong absorption of the wave.

For arbitrary wavelengths, the dispersion equation (12.6) cannot be solved analytically, even if we leave the finite number of terms in the formula (10.18) for the Landau correlation function. However, in some extreme cases, it is possible to obtain simple explicit expressions for the spectra of spin waves.

The spectra of spin waves in a magnetic field directed along the normal to the layers

Consider the spectra of spin waves in some special cases, when the conductors are placed in a magnetic field $\mathbf{B}_0 = (0, 0, B_0)$ directed along the normal to the layers. Let us use model (7.3) for energy distribution of conduction electrons. If the condition of weak spatial dispersion in the plane of the layers $k_x v_F \ll \omega_B$ is met, the dispersion equation is considerably simplified, and simple analytic expressions for the frequencies at arbitrary values of $\eta k_z v_F$ can be obtained. Expanding the exponential function in Eq. (12.3) in powers of $k_x v_F/\omega_B$, we obtain, after simple manipulations, that the matrix $\langle g_{np}\rangle_\beta$ is diagonal

$$\langle g_{np}\rangle_\beta \equiv \langle g_n(\omega, k_x, k_z)\rangle_\beta \delta_{np} = \delta_{np}\left\{\bar{g}_n(\omega, 0, k_z) - (k_x r_0)^2 \sum_{m=-1,0,1} (-1)^m \times \frac{(1+\delta_{m0})}{4} \bar{g}_{n+m}(\omega, 0, k_z)\right\}, \tag{12.8}$$

where

$$\bar{g}_n(\omega, 0, k_z) = \frac{\omega_B \operatorname{sign}(\omega - n\omega_B - \Omega_s)}{\sqrt{(\omega - n\omega_B - \Omega_s)^2 - (\eta k v_1)^2}},$$

and the set of Eqs. (12.2) splits into a set of independent equations. The homogeneous integral equation for the "free" spin-density oscillations has solutions of the form $\Phi_n(p_B, \varphi) = \tilde{\Phi}_n(p_B)e^{in\varphi}$. The frequencies ω_n corresponding to $\Phi_n(p_B, \varphi)$ are determined by the following relationship:

$$1 - \lambda_n \frac{\omega_n}{\omega_B} \bar{g}_n(\omega_n, \mathbf{k}) = 0. \tag{12.9}$$

If the wave propagates in a longitudinal direction, it is more convenient to use Eq. (10.33), which takes the following form for

$\mathbf{k} = (0, 0, k_z)$:

$$\Phi^{(+)} = -\frac{\mu_0 B_+^{\sim}(k_z v_z + \Omega_s)}{\tilde{\omega} - k_z v_z - \Omega_s} + \omega \sum_{p=-\infty}^{\infty} \lambda_p \frac{\bar{\Phi}_p^{(+)} e^{ip\varphi}}{\tilde{\omega} - k_z v_z - p\omega_B - \Omega_s}, \tag{12.10}$$

Integrating (12.10) on $d\varphi$ and $d\beta$, we obtain the algebraic equation for $\bar{\Phi}_0^{(+)}$.

$$\bar{\Phi}_0^{(+)} = \mu_0 B_+^{\sim} \left\langle \frac{kv_z + \Omega_s}{\tilde{\omega} - kv_z - \Omega_s} \right\rangle_\beta + \lambda_0 \bar{\Phi}_0^{(+)} \left\langle \frac{\omega}{\tilde{\omega} - kv_z - \Omega_s} \right\rangle_\beta \tag{12.11}$$

Using this equation to determine $\langle \bar{\Phi}_0^{(+)} \rangle_{\beta,\varphi}$, we obtain the high-frequency magnetic susceptibility:

$$\chi_+(\omega, \mathbf{k}) = \chi_0 \frac{\sqrt{(\omega - \Omega_s)^2 - (\eta k_z v_F)^2} - \omega \operatorname{sign}(\omega - \Omega_s)}{\sqrt{(\omega - \Omega_s)^2 - (\eta k_z v_F)^2} - \lambda_0 \omega \operatorname{sign}(\omega - \Omega_s)}. \tag{12.12}$$

The frequency of the spin wave is determined by the following expression

$$\omega = \frac{\Omega + \sqrt{\lambda_0^2 \Omega^2 + (\eta k_z v_F)^2 (1 - \lambda_0^2)}}{(1 - \lambda_0^2)}. \tag{12.13}$$

In the limiting case of strong spatial dispersion $\eta k_z v_F \gg \Omega_s$, we obtain

$$\omega = \frac{\eta k_z v_1}{\sqrt{1 - \lambda_0^2}}.$$

In the long-wavelength limit $\eta k_z v_F \ll \Omega_s$, the frequency (12.13) coincides with the spin-resonance frequency ω_s for free electrons.

Under the conditions of weak spatial dispersion in the x direction, the frequency shift in Eq. (12.13) proportional to $(k_x v_F/\omega_B)^2$, can easily be found from Eq. (12.9)

$$\Delta\omega = \frac{\lambda_0^3 \omega^3}{\omega_B \sqrt{\lambda_0^2 \Omega_s^2 + (\eta k_z v_1)^2 (1 - \lambda_0^2)}} \Delta\bar{g}(\omega, \mathbf{k}), \tag{12.14}$$

where ω is determined by Eq. (12.13), and $\Delta\bar{g}(\omega, \mathbf{k}) = \bar{g}_0(\omega, k_x, k_z) - \bar{g}_0(\omega, 0, k_z)$.

Spin waves under strong spatial dispersion.
In degenerate conductors with a quasi-two-dimensional charge carrier energy spectrum, the domain of existence of spin waves is broader than that in metals [133]. In layered conductors with metal-type conductivity, the propagation of spin waves for any direction of the wave vector is possible even under conditions of strong spatial dispersion. This is because the drift velocity of conduction electrons is a quantity of the second order of smallness in low-dimensionality parameter η of the electron energy spectrum at certain orientations of the magnetic field deflected significantly from the layers. As a result, there is no Landau damping and the spin density oscillation decays in the relaxation time in the system of electrons and their spins, i.e., it is undamped in the collisionless limit.

Under strong spatial dispersion $\mathbf{kv}_F \gg \omega_B$, Ω_s, the integrals with respect to φ, and φ_1 in formulas (12.3) can be calculated by the stationary phase method. The formula for coefficients g_{np}, defining spectra of high-frequency modes with frequencies ω up to k_{vF} has the form

$$g_{np} = \frac{1}{i}\left\{1 - \exp\left(2\pi i \frac{\tilde{\omega} - \langle \mathbf{kv} \rangle_\varphi}{\omega_B}\right)\right\}^{-1} \times \sum_\alpha \kappa(\boldsymbol{\varphi}^{(\alpha)}) \frac{\exp\left[i(p-n)\varphi^{(\alpha)} - ip\varphi_1^{(\alpha)} + iG(\varphi^{(\alpha)}, \varphi_1^{(\alpha)}) + i\frac{\pi}{4}s\right]}{\sqrt{|\det(G''_{\varphi\varphi_1}(\varphi^{(\alpha)}, \varphi_1^{(\alpha)}))|}} \tag{12.15}$$

Here

$$G(\varphi, \varphi_1) = \frac{\tilde{\omega} \mp \Omega_s}{\omega_B}\varphi_1 - \frac{1}{\omega_B}\int_{\varphi-\varphi_1}^{\varphi} d\varphi' \mathbf{kv}(\varphi'), \quad k_x v_F > \omega.$$

The summation in formula (12.15) is carried out over all the stationary points $\boldsymbol{\varphi}^{(\alpha)} = (\varphi^{(\alpha)}, \varphi_1^{(\alpha)})$, determined from the equations

$$v_x(\varphi) = \frac{\tilde{\omega} \mp \Omega_s}{\omega_B}, \quad v_x(\varphi - \varphi_1) = \frac{\tilde{\omega} \mp \Omega_s}{\omega_B}.$$

$\kappa(\boldsymbol{\varphi}^{(\alpha)}) = 1$ for the stationary points within the integration area $0 < \varphi^{(\alpha)} < 2\pi$, $0 < \varphi_1^{(\alpha)} < 2\pi$ and $\kappa(\boldsymbol{\varphi}^{(\alpha)}) = 1/2$ for the points located on its boundary,

$$s = \mathrm{sign} G''_{\varphi\varphi_1}(\varphi^{(\alpha)}, \varphi_1^{(\alpha)}) = \nu_+(R''_{\varphi\varphi_1}) - \nu_-(R''_{\varphi\varphi_1})$$

where $\nu_+(R''_{\varphi\varphi_1})$ and $\nu_-(R''_{\varphi\varphi_1})$ are the number of positive and negative eigenvalues of the matrix

$$G''_{\varphi\varphi_1} \equiv \frac{\partial^2 G(\varphi^{(\alpha)}, \varphi_1^{(\alpha)})}{\partial\varphi\partial\varphi_1}, \text{respectively.}$$

For the directions of $\mathbf{B}_0 = (B_0 \sin\vartheta, 0, B_0 \cos\vartheta)$ relative to the conductor layers such that $\mathbf{v}_D$ is close to zero, wave attenuation is defined by relaxation processes in the system of electrons and their spins and existence of collective modes is possible even at $\eta k v_F \simeq \omega_B$. In this case, there are solutions of the dispersion equation (12.6) in the vicinity of the resonance

$$\omega = n_1\omega_B \pm \Omega_s + \Delta\omega, \quad |\Delta\omega| \ll \omega_B, \quad n_1 = 0, 1, 2\ldots. \tag{12.16}$$

If wave the vector $\mathbf{k} = (k\sin\phi, 0, k\cos\phi)$ is oriented in the xz plane, the correction to resonance frequency can be written as

$$\Delta\omega = \frac{n_1\omega_B \pm \Omega_s}{\pi k_x r_0}\gamma_i, \tag{12.17}$$

where $r_0 = v_F/\omega_B$ is the Larmor radius of the conduction electron, and γ_i are roots of the algebraic equation

$$\det|\delta_{np} - \lambda_p \gamma_i^{-1} \langle I_{np}(\beta)\rangle_\beta| = 0, \tag{12.18}$$

$$I_{np}(\beta) = \sum_\alpha \kappa(\boldsymbol{\varphi}^{(\boldsymbol{\alpha})}) \frac{\exp\left[i(p-n)\varphi^{(\alpha)} - ip\varphi_1^{(\alpha)} + iG_1(\varphi^{(\alpha)}, \varphi_1^{(\alpha)}) + i\frac{\pi}{4}s\right]}{\sqrt{|\det(G''_{1\,\varphi\varphi_1}(\varphi^{(\alpha)}, \varphi_1^{(\alpha)}))|}}$$

here $G_1(\varphi^{(\alpha)}, \varphi_1^{(\alpha)}) = G(\varphi^{(\alpha)}, \varphi_1^{(\alpha)})|_{\omega = n_1\omega_B \pm \Omega_s}$.

Assuming that the energy of conduction electrons and velocity components are defined by the formulas (7.3), and (7.4)–(7.6), respectively, one can obtain, at $kr_0 \gg 1$, $\omega < kv_F$, following asymptotic expression for the coefficients $g_{np}(\beta)$, [134]

$$g_{np}(\beta) = \frac{1}{k_x r_0(1-\rho^2)} \left\{ \cos[(n-p)\delta] \cot\left(\pi \frac{\langle \tilde{\omega} \mp \Omega - \mathbf{kv} \rangle_\varphi}{\omega_B(\beta)}\right) + \sin\left[\frac{1}{\omega_B(\beta)} \int_{-\delta}^{\delta} d\varphi \mathbf{kv} + \pi \frac{\tilde{\omega} \mp \Omega - \langle \mathbf{kv} \rangle_\varphi}{\omega_B(\beta)} - 2\delta \frac{(\tilde{\omega} \mp \Omega)}{\omega_B(\beta)} + (n+P)\delta \right] \times \sin^{-1}\left(\pi \frac{\tilde{\omega} \mp \Omega - \langle \mathbf{kv} \rangle_\varphi}{\omega_B(\beta)}\right) \right\}. \tag{12.19}$$

Here $\delta = \arccos(\tilde{\omega} \mp \Omega)/\omega_B k_x r_0$, $\rho = (\tilde{\omega} \mp \Omega)/\omega_B k_x r_0$, $\beta = p_B/p_0 \cos\vartheta$.

For not too large n and p, and $\omega \ll k v_F$, formula (12.19) can be transformed to the form

$$g_{np}(\beta) = \frac{1}{k_x r_0} \left(\cot \frac{\pi(\tilde{\omega} \mp \Omega_s)}{\omega_B} \cos\frac{\pi}{2}(n-p) + \frac{\sin\left(R(\vartheta_i) + \frac{\pi}{2}(n+p)\right)}{\sin \frac{\pi(\tilde{\omega} \mp \Omega_s)}{\omega_B}} \right) \tag{12.20}$$

and Eq. (12.18) is simplified

$$\det\left| \delta_{np} - \lambda_p \gamma_i^{-1} \left(\cos\frac{\pi}{2}(n-p) + (-1)^{n_1} \left\langle \sin\left(R(\vartheta_i) + \frac{\pi}{2}(n+p)\right) \right\rangle_\beta \right) \right| = 0. \tag{12.21}$$

Here

$$R(\vartheta_i) = \int_{-\pi/2}^{\pi/2} \frac{\mathbf{kv}(\varphi)}{\omega_B(\beta_i)} d\varphi,$$

ϑ_i is roots of the equation $J_0((m v_F/p_0)\tan\vartheta_i) = 0$.

In the magnetic field parallel to the normal to layers $\mathbf{B}_0 = (0, 0, B_0)$ formulas (12.20) and (12.21) take the form

$$\bar{g}_{np} = \frac{1}{k_x r_0}\left(\cot\frac{\pi(\tilde{\omega} - \Omega_s)}{\omega_B}\cos\frac{\pi}{2}(n-p) + \mathrm{J}_0(\varepsilon k_x r_0) \times \frac{\sin\left(2k_x r_0 + \frac{\pi}{2}(n+p)\right)}{\sin\frac{\pi(\tilde{\omega} - \Omega_s)}{\omega_B}}\right), \tag{12.22}$$

$$\det\left[\delta_{np} - \lambda_p \gamma^{-1}\left\{\cos\frac{\pi}{2}(n-p) + (-1)^m \mathrm{J}_0(\varepsilon k_x r_0) \times \sin\left(2k_x r_0 + \frac{\pi}{2}(n+p)\right)\right\}\right] \tag{12.23}$$

where $\varepsilon = 2\eta v_F p_0/\varepsilon_F$.

In the model determining the correlation function by the zero and first Fourier harmonics

$$S(\mathbf{p}, \mathbf{p}') = S_0 + 2S_1 \cos(\varphi - \varphi'), \tag{12.24}$$

equation (12.21) is reduced to the quadratic equation the roots of which are equal

$$\gamma_{1,2} = \frac{1}{2}(\lambda_0 + 2\lambda_1 + (-1)^{n_1}(\lambda_0 - 2\lambda_1)g \pm \{(\lambda_0 + 2\lambda_1 + (-1)^{n_1}(\lambda_0 - 2\lambda_1)g)^2 + 8\lambda_0\lambda_1(-1 + g^2 + h^2)\}^{1/2} \tag{12.25}$$

where $g = \langle \sin R(\vartheta)\rangle_\beta$, $h = \langle \cos R(\vartheta)\rangle_\beta$. When the correlation function is defined by Eq. (12.24) and the condition $\varepsilon k_x r_0 \ll 1$ is satisfied (at the same time condition $k_x r_0 \gg 1$ remains), the asymptotics of Eq. (12.23) has the only one root

$$\gamma = \lambda_0 + 2\lambda_1 + (-1)^m(\lambda_0 - 2\lambda_1)\sin(2k_x r_0). \tag{12.26}$$

As is seen from Eqs. (12.17), (12.25), (12.26) in the vicinity of the resonance, frequency is quickly oscillating function k_x. Let us note that because of the smallness of η, the inequality $k_z v_z \simeq \eta k_z v_F \ll \Omega_s$,

ω_B is carried out in the wide range of values k_z, even in not really strong magnetic fields.

At $\omega - \Omega_s > kv_F$, the rapidly oscillating phase of the exponent in Eq. (12.3) does not have stationary points, and the asymptotic behavior of (12.3) is calculated by the integration by parts. Substituting the result in (12.6) we obtain following asymptotic expression for the spectrum of high-frequency ($\omega \gg \omega_B$) spin mode

$$\omega = \gamma_i k v_F, \qquad (12.27)$$

where γ_i are the roots of equation

$$\det\left[\delta_{np} - \lambda_p \left\langle \frac{e^{i(n-p)\varphi}}{1 - \gamma^{-1}\mathbf{kv}(\beta,\varphi)/(kv_F)} \right\rangle_{\beta,\varphi}\right] = 0,$$

Real constants γ_i greater than unity correspond to the wave processes.

Spin waves can be detected experimentally by monitoring the selective transparency of thin films in the vicinity of the frequencies of spin paramagnetic resonance. The comparison of the values of $\omega(\mathbf{k})$ for different values of wave numder would make it possible to determine the constant which characterizes the exchange interaction between the charge carriers.

Conclusion

In this review we have concentrated on the phenomena in strongly anisotropic organic conductors in normal metallic state that arise due to the dynamics of conduction electrons in strong magnetic fields. The magnetotransport and high-frequency resonance effects contain important information about the parameters and topological structure of the FS, Fermi-liquid interaction between charge carriers and relaxation processes. The presenting results may be suitable for the analysis of experimental data in other layered structures such as dichalcogenides of transition metals, cuprates, etc., possessing the metal type of electrical conductivity and sharp anisotropy of electron energy spectrum.

We are grateful to M.V. Kartsovnik for fruitful discussions of the problem under study.

List of references

1. F. Schegolev, Phys. Status Solidi A **12**, 9 (1972).
2. D. Jerome and H. J. Schulz, Adv. Phys. **31**, 299 (1982).
3. T. Ishiguro, K. Yamaji, and G. Saito, Organic Superconductors (Springer Verlag, Berlin, 1998).
4. E. B. Yagubskii, I. F. Shchegolev, V. N. Laukhin, P. A. Kononovich, M. V. Karatsovnik, A. V. Zvarykina, and L. I. Buravov, Pis'ma Zh. Eksp. Teor. Fiz. **39**, 12 (1984) [JETP Lett. **39**, 12 (1984)].
5. M. V. Kartsovnik, V. N. Laukhin, V. I. Nizhankovskii, and A. A. Ignat'ev, Pis'ma Zh. Eksp. Teor. Fiz. **47**, 302 (1988) [JETP Lett. **47**, 363 (1988)].
6. M. V. Kartsovnik, P. A. Kononovich, V. N. Laukhin, and I. F. Shchegolev, Pis'ma Zh. Eksp. Teor. Fiz. **48**, 498 (1988) [JETP Lett. **48**, 541 (1988)].
7. D. Parker, D. D. Pigram, R. H. Friend, M. Kurmo, and P. Day, Synth. Met. **27**, A387 (1988).
8. K. Oshima, T. Mori, H. Inokuchi, H. Urayama, H. Yamochi, and C. Saito, Phys. Rev. B **38**, 938(R) (1988).
9. N. Toyota, T. Sasaki, T. Murata, Y. Honda, M. Tokumoto, H. Bando, N. Kinoshima, and H. Anzai, J. Phys. Soc. Jpn. **57**, 2616 (1988).
10. K. Kajita, Y. Nishio, T. Takahashi, W. Sasaki, R. Kato, and H. Kobayashi, Solid State Commun. **70**, 1189 (1989).
11. W. Kang, G. Montambaux, J. R. Cooper, D. Jerome, P. Batai, and C. Lenoir, Phys. Rev. Lett. **62**, 2559 (1989).
12. N. Thorup, G. Rindorf, H. Soling and K. Bechgaard, Acta Cryst. **B37**, 1236 (1981).
13. H. Urayama, H. Yamochi, G. Saito, S. Sato, A. Kawamoto, J. Tanaka, T. Mori, Y. Maruyama, and H. Inokuchi, Chem. Lett. **17**, 463 (1988).
14. V. F. Kaminskii, T. G. Prokhorova, R. P. Shibaeva, and E. B. Yagubskii, Pis'ma Zh. Eksp. Teor. Fiz. **39**, 15 (1984) [JETP Lett. **39**, 17 (1984)].
15. H. Kobayashi, R. Kato, A. Kobayashi, G. Saito, M. Tokumoto, H. Anzai, and T. Ishiguro, Chem. Lett. **14**, 1293 (1985).
16. H. Kuroda, K. Yakushi, H. Tajima, A. Ugawa, Y. Okawa, A. Kobayashi, R. Kato, H. Kobayashi, and G. Saito, Synth. Met. **27**, A491 (1988).
17. K. Yamaji, J. Phys. Soc. Jpn. **58**, 1520 (1989).
18. V. G. Peschansky, J. A. Roldan Lopez, and T. G. Yao, J. Phys. Fr. **1**, 1469 (1991).
19. O. V. Kirichenko and V. G. Peschansky, Fiz. Nizk. Temp. **37**, 925 (2011) [Low Temp. Phys. **37**, 734 (2011)].
20. M. Lifshitz, Zh. Eksp. Teor. Fiz. **38**, 1569 (1960) [Sov. Phys. JETP **11**, 1130 (1960)].
21. R. Kubo, J. Phys. Soc. Jpn. **12**, 570 (1957).
22. A. I. Akhiezer, S. V. Peletminsky, Methods of the statistical physics, Nauka, Moskow (1977) [in Russian].
23. O. V. Konstantinov and V. I. Perel', Zh. Eksp. Teor. Fiz. **37**, 786 (1959)[Sov. Phys. JETP **10**, 560 (1960)].
24. V. G. Peschansky, and D. I. Stepanenko, Fiz. Nizk. Temp. **33**, 591 (2007) [Low Temp. Phys. **33**, 443 (2007)].
25. M. Lifshitz and V. G. Peschanskii, Zh. Eksp. Teor. Fiz. **35**, 1251 (1958) [Sov. Phys. JETP **8**, 875 (1959)].
26. P. Kapitza, Proc. R. Soc. A **129**, 358 (1928).
27. Hasan R. Atalla, Low Tem. Phys. **29**, 593 (2003).
28. M. V. Kartsovnik and V. N. Laukhin, J. Phys. Fr. **6**, 1753 (1996).

29. J. Wosnitza, Fermi Surface of Low-Dimensional Organic Metals and Superconductors, Springer Tracts in Modern Physics (Springer Verlag, Berlin, 1996) **134**, 1 (1996).
30. V. G. Peschansky, Phys. Rep. **288**, 305 (1997).
31. J. Singleton, Rep. Prog. Phys. **63**, 1111 (2000).
32. M. V. Kartsovnik, Chem. Rev. **104**, 5737 (2004).
33. M. V. Kartsovnik and V. G. Peschansky, Fiz. Nizk. Temp. **31**, 249 (2005) [Low. Temp. Phys. **31**, 185 (2005)].
34. M. V. Kartsovnik, in The Physics of Organic Superconductors and Conductors, edited by A. G. Lebed, Springer Series in Material Sciences (Springer Verlag, Berlin, Heidelberg, 2008), **110**, 185 (2008).
35. V. G. Peschansky, D. I. Stepanenko, Fiz. Nizkh. Temp. **42**, 1221 (2016). [Low Temp. Phys. **42**, 947 (2016)]
36. Carrington, Rep. Prog. Phys. **74**, 124507 (2011).
37. F. Schegolev, P. A. Kononovich, V. N. Laukhin, and M. V. Kartsovnik, Phys. Scr. **29**, 46 (1989).
38. R. Yagi, Y. Iye, Y. Hashimoto, T. Odagiri, H. Noguchi, H. Sasaki, and T. Ikoma, J. Phys. Soc. Jpn. **60**, 3784 (1991).
39. Y. Iye, M. Baxendale, and V. Z. Mordkovich, J. Phys. Soc. Jpn. **63**, 1643 (1994).
40. M. Baxendale, V. Z. Mordkovich, and S. Yoshimura, Solid State Commun. **107**, 165 (1998).
41. E. Ohmichi, H. Adachi, Y. Mori, Y. Maeno, T. Ishiguro, and T. Oguchi, Phys. Rev. B **59**, 7263 (1999).
42. V. V. Eremenko, V. A. Sirenko, in Physics Reviews **23**, Cambridge Scientific Publishers (2007).
43. G. M. Danner, W. Kang, and P. M. Chaikin, Phys. Rev. Lett. 72, 3714 (1994).
44. T. Osada, A. Kawasumi, S. Kagoshima, N. Miura, and G. Saito, Phys. Rev. Lett. **66**, 1525 (1991).
45. M. J. Naughton, O. H. Chung, M. Chaparala, X. Bu, and P. Coppens, Phys. Rev. Lett. **67**, 3712 (1991).
46. T. Osada, S. Kagoshima, and N. Miura, Phys. Rev. B **46**, 1812 (1992).
47. G. Lebed, Pis'ma Zh. Eksp. Teor. Fiz. **43**, 137 (1986) [JETP Lett. **43**, 174 (1986)].
48. L. D. Landau, Z. Phys. 64, 629 (1930).
49. L. W. Shubnikov and W. J. de Haas, Leiden Commun. **207**, 210 (1930).
50. W. J. de Haas and P. M. van Alphen, Leiden Commun. A **212**, 215 (1930).
51. L. Onsager, Philos. Mag. **43**, 1006 (1952).
52. M. Lifshitz and A. M. Kosevich, Zh. Eksp. Teor. Fiz. **29**, 730 (1955) [English Translation Ukr. J. Phys. **53**, 112 (2008)].
53. R. B. Dingle, Proc. R. Soc. A **211**, 517 (1952).
54. J. Wosnitza, G. Goll, D. Beckmann, S. Wanka, D. Schweitzer, and W. Strunz, J. Phys. Fr. **6**, 1597 (1996).
55. J. Wosnitza, G. W. Crabtree, J. M. Williams, H. H. Wang, K. D. Carlson, and U. Geiser, Synth. Met. **56**, 2891 (1993).
56. Yu. A. Bogod, V. B. Krasovitskii, and V. G. Gerasimenko, Zh. Eksp. Teor. Fiz. **66**, 1362 (1974) [Sov. Phys. JETP **39**, 667 (1974)].
57. Yu. A. Bogod, V. B. Krasovitskii, and S. A. Mironov, Zh. Eksp. Teor. Fiz. **78**, 1099 (1980).
58. Yu. A. Bogod, Vit. B. Krasovitsky, and E. T. Lemesevskaya, Fiz. Nizk. Temp. **9**, 832 (1983) [Low Temp. Phys. **9**, 431 (1983)].
59. Yu. A. Bogod, Vit. B. Krasovitsky, and E. T. Lemesevskaya, Fiz. Nizk. Temp. **12**, 610 (1986) [Low Temp. Phys. **12**, 345 (1986)].
60. V. M. Polyanovskii, Pis'ma Zh. Eksp. Teor. Fiz. **46**, 108 (1987) [JETP Lett. **46**, 132 (1987)].

61. V. Polyanovsky, Phys. Rev. B **47**, 1985 (1993).
62. M. V. Kartsovnik, P. D. Grigoriev, W. Biberacher, N. D. Kushch, and P. Wyder, Phys. Rev. Lett. **89**, 126802 (2002).
63. M. V. Kartsovnik, P. A. Kononovich, V. N. Laukhin, S. I. Pesotskii, and I. F. Shchegolev, Pis'ma Zh. Eksp. Teor. Fiz. **49**, 453 (1989) [JETP Lett. **49**, 520 (1989)].
64. E. Ohmichi, H. Ito, T. Ishiguro, G. Saito, and T. Komatsu, Phys. Rev. B **57**, 7481 (1998).
65. T. G. Togonidze, M. V. Kartsovnik, J. A. A. J. Perenboom, N. D. Kushch, and H. Kobayashi, Physica B **294–295**, 435 (2001).
66. L. Balicas, J. S. Brooks, K. Storr, D. Graf, S. Uji, H. Shinagawa, E. Ojima, H. Fujiwara, H. Kobayashi, A. Kobayashi, and M. Tokumoto, Solid State Commun. **116**, 557 (2000).
67. L. van Hove, Phys. Rev. **89**, 1189 (1953).
68. A. Varlamov, V. S. Egorov, and A. V. Pantsulaya, Adv. Phys. **38**, 469 (1989).
69. D. Shoenberg – Report Satelit Conference "Fermi Surface in Metals" to Intern. Conf. Low Temp. Phys. LT7, Toronto, 1960, (edit. John Wiley and Sons, New York, 1960) p. 80.
70. M. H. Cohen and L. M. Falicov, Phys. Rev. Lett. **7**, 231 (1961)
71. A. B. Pippard, Proc. Roy. Soc. **A270**, 1 (1962),
72. L. M. Falicov, A. B. Pippard, P. R. Sieverd, Phys. Rev., **151**, 498 (1966).
73. V. G. Peschanskii, Zh. Eksp. Teor. Fiz. **52**, 1312 (1967) [JETP **25**, 872 (1967)].
74. D. Shoenberg, "Magnetic oscillations in metals", Cambridge Univ. Press., Cambridge 1984.
75. G. G. Lonzarich, Electrons at the Fermi Surface (ed. M. Springford), Cambridge Univ. Press, Cambridge 1980.
76. G. G. Lonzarich and P. M. Holtham, Proc. Roy Soc. London **A400,** 145 (1986).
77. K. Andres, C.-P. Heidmann, H. Müller, S. Himmellsbach, W. Biberacher, Ch. Probst, W. Joss, Synth. Metals **41-43**, 1893 (1991).
78. C.-P. Heidmann, H. Müller, W. Biberacher, K. Neumaier, Ch. Probst, K. Andres, A. G. M. Jansen, W. Joss, Synth. Metals **41-43**, 2029 (1991).
79. A. F. Bangura, P. A. Coddard, J. Singelton, S. W. Tozer, A. I. Coldea, A. Ardovan, R. D. McDonald, S. J. Blundell, and J. A. Schlueter, Phys. Rev. B **76**, 052510 (2007).
80. M. V. Kartsovnik, V. N. Zverev, D. Andres, W. Biberacher, T. Helm, P. D. Grigoriev, R. Ramazashvili, N. D. Kushch, and H. Müller, Fiz. Nizk. Temp. **40**, 484 (2014) [Low Temp. Phys. **40**, 377 (2014)].
81. T. Helm, M. V. Kartsovnik, I. Sheikin, M. Bartkowiak, F. Wolff-Fabris, N. Bittner, W. Biberacher, M. Lambacher, A. Erb, J. Wosnitza, and R. Gross, Phys. Rev. Lett. **105**, 247002 (2010).
82. M. V. Kartsovnik, T. Helm, C. Putzke, F. Wolff-Fabris, I. Sheikin, S. Lepault, C. Proust, D. Vignolles, N. Bittner, and W. Biberacher, New J. Phys. **13**, 015001 (2011).
83. T. Helm, M. V. Kartsovnik, C. Proust, B. Vignolle, C. Putzke, E. Kampert, I. Sheikin, E.-S. Choi, J. S. Brooks, N. Bittner, W. Biberacher, A. Erb, J. Wosnitza, and R. Gross, Phys. Rev. B **92**, 094501 (2015).
84. O. Galbova, O. V. Kirichenko, and V. G. Peschansky, Fiz. Nizk. Temp. **39**, 780 (2013) [Low Temp. Phys. **39**, 602 (2013)].
85. V. G. Peschansky and D. I. Stepanenko, Zh. Eksp. Teor. Fiz. **150**, 176 (2016) [JETP **123**, 156 (2016)].
86. V. G. Peschansky, Fiz. Nizk. Temp. **43,** 303 (2017) [Low Temp. Phys. **43**, 602 (2017)].
87. E. J. Blount, Phys. Rev. **126,** 1636 (1962).

88. M. G. Priestley, Proc.Roy.Soc. **A276**, 258 (1963).
89. O. Galbova, V. G. Peschansky, and D. I. Stepanenko, Int. J. Mod. Phys. B **31**, 1750114 (2017)
90. G. E. Volovik Fiz. Nizk. Temp, **43**, 57 (2017) [Low Temp. Phys. **43**, 47 (2017)].
91. V. G. Peschansky, Fiz. Nizk. Temp. **23**, 47 (1997) [Low Temp. Phys. **23**, 42 (1997)].
92. V. G. Peschansky, Zh. Eksp. Teor. Fiz. **121**, 1204 (2002) [JETP **94**, 1035 (2002)].
93. O. Galbova, V. G. Peschansky, and D. I. Stepanenko, Fiz. Nizk. Temp. **41**, 691 (2015) [Low Temp. Phys. **41**, 537 (2015)].
94. J. Y. Fortin and J. Audouard, Fiz. Nizk. Temp. **43,** 211(2017) [Low Temp. Phys. **43**, 173 (2017)].
95. M. I. Azbel' and E. A. Kaner, Zh. Eksp. Teor. Fiz. **32**, 896 (1957) [Sov. Phys. JETP **5**, 730 (1957)].
96. J. Singleton, F. L. Pratt, M. Doporto, T. J. B. M. Janssen, M. Kurmoo, J. A. A. J. Perenboom, W. Hayes, and P. Day, Phys. Rev. Lett. **68**, 2500 (1992).
97. S. Hill, A. Wittlin, J. van Bentum, J. Singleton, W. Hayes, J. A. A. J. Perenboom, M. Kurmoo, and P. Day, Synth. Met. **70**, 821 (1995).
98. S. Hill, J. Singleton, F. L. Pratt, M. Doporto, W. Hayes, T. J. B. M. Janssen, J. A. A. J. Perenboom, M. Kurmoo, and P. Day, Synth. Met. **56**, 2566 (1993).
99. J. Singleton, F. L. Pratt, M. Doporto, J. M. Caulfield, S. O. Hill, T. J. B. M. Janssen, I. Deckers, G. Pitsi, F. Herlach, W. Hayes, J. A. A. J. Perenboom, M. Kurmoo, and P. Day, Physica B **184**, 470 (1993).
100. S. V. Demishev, N. E. Sluchanko, A. V. Semeno, and N. A. Samarin, Pis'ma Zh. Eksp. Teor. Fiz. **61**, 299 (1995) [JETP Lett. **61**, 313 (1995)].
101. S. V. Demishev, A. V. Semeno, N. E. Sluchanko, N. A. Samarin, I. B. Voskoboinikov, V. V. Glushkov, A. E. Kovalev, and N. D. Kusch, Pis'ma Zh. Eksp. Teor. Fiz. **62**, 215 (1995) [JETP. Lett. **62**, 228 (1995)].
102. S. V. Demishev, A. V. Semeno, N. E. Sluchanko, N. A. Samarin, I. B. Voskoboinikov, V. V. Glushkov, J. Singleton, S. J. Blundell, S. O. Hill, W. Hayes, M. V. Kartsovnik, A. E. Kovalev, M. Kurmoo, P. Day, and N. D. Kushch, Phys. Rev. B **53**, 12794 (1996).
103. S. V. Demishev, A. V. Semeno, N. E. Sluchanko, N. A. Samarin, I. B. Voskoboinikov, M. V. Kartsovnik, A. E. Kovalev, and N. D. Kushch, Zh. Eksp. Teor. Fiz. **111**, 979 (1997) [JETP **84**, 540 (1997)].
104. Y. Oshima, H. Ohta, K. Koyama, M. Motokawa, H. M. Yamamoto, and R. Kato, J. Phys. Soc. Jpn. **71**, 1031 (2002).
105. Y. Oshima, H. I. Ohta, K. Koyama, M. Motokawa, H. M. Yamamoto, R. Kato, M. Tamura, Y. Nishio, and K. Kajita, J. Phys. Soc. Jpn. **72**, 143 (2003).
106. E. Kovalev, S. Hill, K. Kawano, M. Tamura, T. Naito, and H. Kobayashi, Phys. Rev. Lett. **91**, 216402 (2003).
107. S. Hill, Phys. Rev. B **55**, 4931 (1997).
108. H. Ohta, M. Kimata, and Y. Oshima, Sci. Technol. Adv. Mater. **10**, 024310 (2009).
109. A.Ardavan, J. M. Schrama, S. J. Blundell, J. Singleton, W. Hayes, M. Kurmoo, P. Day, and P. Goy, Phys. Rev. Lett. **81**, 713 (1998).
110. E. Kovalev, S. Hill, and J. S. Quall, Phys. Rev. B **66**, 134513 (2002).
111. Y. Oshima, M. Kimata, K. Kishigi, H. Ohta, K. Koyama, M. Motokawa, H. Nishikawa, K. Kikuchi, and I. Ikemoto, Phys. Rev. B **68**, 054526 (2003).
112. Y. Oshima, M. Kimata, K. Kishigi, H. Ohta, K. Koyama, M. Motokawa, H. Nishikawa, K. Kikuchi, and I. Ikemoto, Physica B **346–347**, 387 (2004).
113. S. Takahashi, S. Hill, S. Takasaki, J. Yamada, and H. Anzai, Phys. Rev. B **72**, 024540 (2005).
114. F. J. Dyson, Phys. Rev. **98**, 349 (1955).

115. V. G. Peschansky and V. S. Lekhtsier, Zh. Eksp. Teor. Fiz. **46**, 764 (1964) [JTEP **19**, 520 (1964)].
116. M. I. Azbel and V. G. Peschansky, Zh. Eksp. Teor. Fiz. **54**, 477 (1968) [JTEP **27**, 260 (1968)].
117. D. I. Stepanenko, Solid State Commun. **150**, 1204 (2010).
118. M. V. Fedoryuk, Asymptotics: Integrals and Series (Nauka, Moscow, 1987) [in Russian].
119. V. G. Peschansky and D. I. Stepanenko, Fiz. Nizk. Temp. **40**, 851 (2014) [Low Temp. Phys. **40**, 662 (2014)].
120. O. V. Kirichenko, V. G. Peschansky and D. I. Stepanenko, Zh. Eksp. Teor. Fiz. **126**, 1435 (2004) [JETP **99**, 1253 (2004)].
121. M. Ya. Azbel, Zh. Eksp. Teor. Fiz. **39**, 400 (1960) [Sov. Phys. JETP **12**, 283 (1961)].
122. P. M. Platzman and P. A. Wolf, Waves and Interactions in Solid State Plasmas (Academic Press, New York, 1973).
123. E. A. Kaner and V. G. Skobov, Adv. Phys. **17**, 605 (1968).
124. O. V. Kirichenko, V. G. Peschansky, and D. I. Stepanenko, Phys. Rev. B **71**, 045304 (2005).
125. D. I. Stepanenko, Modern Phys. Lett. B 26, 1250190 (2012).
126. Yu. A. Kolesnichenko, V. G. Peschansky, D. I. Stepanenko, Fiz. Nizkh. Temp. **43**, 227 (2017). [Low Temp. Phys. **43**, 186 (2017)].
127. D. I. Stepanenko, Fiz. Nizkh. Temp. **44**, 1004 (2018). [Low Temp. Phys. **44**, 786 (2018)].
128. L. D. Landau, Zh. Eksp. Teor. Fiz. **30,** 1058 (1956) [Sov. Phys. JETP **3**, 920 (1956)].
129. V. P. Silin, Zh. Eksp. Teor. Fiz. **33**, 495, (1957) [Sov. Phys. JETP **6**, 387 (1958)].
130. V. P. Silin, Zh. Eksp. Teor. Fiz. **33**, 495, (1957) [Sov. Phys. JETP **6**, 387 (1958)]
131. S. Schultz, G. Dunifer, Phys. Rev. Lett. **18**, 283 (1967).
132. V. G. Peschansky, D. I. Stepanenko, Pis'ma Zh. Eksp. Teor. Fiz. **78**, 772 (2003) [JETP Lett. **78**, 322 (2003)].
133. D. I. Stepanenko, Fiz. Nizk. Temp. **31**, 115 (2005) [Low Temp. Phys. **31**, 90 (2005)].
134. V. G. Peschansky, D. I. Stepanenko, Fiz. Nizkh. Temp. **33**, 1027 (2007) [Low Temp. Phys. **33**, 783 (2007)].

www.ingramcontent.com/pod-product-compliance
Lightning Source LLC
Chambersburg PA
CBHW061832220326
41599CB00027B/5264